NUMBERS

Revised Edition

THE HISTORY OF MATHEMATICS

NUMBERS

COMPUTERS, PHILOSOPHERS,
AND THE SEARCH FOR MEANING

Revised Edition

John Tabak, Ph.D.

Facts On File
An Infobase Learning Company

NUMBERS: Computers, Philosophers, and the Search for Meaning, Revised Edition

Copyright © 2011, 2004 by John Tabak, Ph.D.

Facts On File, Inc.
An imprint of Infobase Learning
132 West 31st Street
New York NY 10001

Library of Congress Cataloging-in-Publication Data

Tabak, John
   Numbers: computers, philosophers, and the search for meaning / John Tabak—Rev. ed.
      p. cm.—(History of mathematics)
   Includes bibliographical references and index.
   ISBN 978-0-8160-7940-7 (alk. paper)
   1. Numeration—History. 2. Counting—History. I. Title.
   QA141.T33 2011
   513.2—dc22   2010015830

Text design by David Strelecky
Composition by Hermitage Publishing Services
Illustrations by Dale Williams
Photo research by Elizabeth H. Oakes
Cover printed by Bang Printing, Brainerd, Minn.
Book printed and bound by Bang Printing, Brainerd, Minn.
Date printed: March 2011
Printed in the United States of America

10 9 8 7 6 5 4 3 2 1

To Sandra and James, world travelers, good company

CONTENTS

PREFACE

Of all human activities, mathematics is one of the oldest. Mathematics can be found on the cuneiform tablets of the Mesopotamians, on the papyri of the Egyptians, and in texts from ancient China, the Indian subcontinent, and the indigenous cultures of Central America. Sophisticated mathematical research was carried out in the Middle East for several centuries after the birth of Muhammad, and advanced mathematics has been a hallmark of European culture since the Renaissance. Today, mathematical research is carried out across the world, and it is a remarkable fact that there is no end in sight. The more we learn of mathematics, the faster the pace of discovery.

Contemporary mathematics is often extremely abstract, and the important questions with which mathematicians concern themselves can sometimes be difficult to describe to the interested nonspecialist. Perhaps this is one reason that so many histories of mathematics give so little attention to the last 100 years of discovery—this, despite the fact that the last 100 years have probably been the most productive period in the history of mathematics. One unique feature of this six-volume History of Mathematics is that it covers a significant portion of recent mathematical history as well as the origins. And with the help of in-depth interviews with prominent mathematicians—one for each volume—it is hoped that the reader will develop an appreciation for current work in mathematics as well as an interest in the future of this remarkable subject.

Numbers details the evolution of the concept of number from the simplest counting schemes to the discovery of uncomputable numbers in the latter half of the 20th century. Divided into three parts, this volume first treats numbers from the point of view of computation. The second part details the evolution of the concept of number, a process that took thousands of years and culminated in what every student recognizes as "the real number line," an extremely important and subtle mathematical idea. The third part of this volume concerns the evolution of the concept of the infinite. In particular, it covers Georg Cantor's discovery (or creation, depending on one's point of view) of transfinite numbers and his efforts to place set theory at the heart of modern mathematics. The most important ramifications of Cantor's work, the attempt to axiomatize mathematics carried out by David Hilbert and Bertrand Russell, and the discovery by Kurt Gödel and Alan Turing that there are limitations on what can be learned from the axiomatic method, are also described. The last chapter ends with the discovery of uncomputable numbers, a remarkable consequence of the work of Kurt Gödel and Alan Turing. The book concludes with an interview with Professor Karlis Podnieks, a mathematician of remarkable insights and a broad array of interests.

Probability and Statistics describes subjects that have become central to modern thought. Statistics now lies at the heart of the way that most information is communicated and interpreted. Much of our understanding of economics, science, marketing, and a host of other subjects is expressed in the language of statistics. And for many of us statistical language has become part of everyday discourse. Similarly, probability theory is used to predict everything from the weather to the success of space missions to the value of mortgage-backed securities.

The first half of the volume treats probability beginning with the earliest ideas about chance and the foundational work of Blaise Pascal and Pierre Fermat. In addition to the development of the mathematics of probability, considerable attention is given to the application of probability theory to the study of smallpox and the misapplication of probability to modern finance. More than most branches of mathematics, probability is an applied discipline, and its uses and misuses are important to us all. Statistics is the subject of the second half of the book. Beginning with the earliest examples of statistical thought, which are found in the writings of John Graunt and Edmund Halley, the volume gives special attention to two pioneers of statistical thinking, Karl Pearson and R. A. Fisher, and it describes some especially important uses and misuses of statistics, including the use of statistics in the field of public health, an application of vital interest. The book concludes with an interview with Dr. Michael Stamatelatos, director of the Safety and Assurance Requirements Division in the Office of Safety and Mission Assurance at NASA, on the ways that probability theory, specifically the methodology of probabilistic risk assessment, is used to assess risk and improve reliability.

Geometry discusses one of the oldest of all branches of mathematics. Special attention is given to Greek geometry, which set the standard both for mathematical creativity and rigor for many centuries. So important was Euclidean geometry that it was not until the 19th century that mathematicians became willing to consider the existence of alternative and equally valid geometrical systems. This 19th-century revolution in mathematical, philosophical, and scientific thought is described in some detail, as are some alternatives to Euclidean geometry, including projective geometry, the non-Euclidean geometry of Nikolay Ivanovich Lobachevsky and János Bolyai, the higher (but finite) dimensional geometry of Riemann, infinite-dimensional geometric ideas, and some of the geometrical implications of the theory of relativity. The volume concludes with an interview with Professor Krystyna Kuperberg of Auburn University about her work in geometry and dynamical systems, a branch of mathematics heavily dependent on ideas from geometry. A successful and highly insightful mathematician, she also discusses the role of intuition in her research.

Mathematics is also the language of science, and mathematical methods are an important tool of discovery for scientists in many disciplines. *Mathematics and the Laws of Nature* provides an overview of the ways that mathematical thinking has influenced the evolution of science—especially the use of deductive reasoning in the development of physics, chemistry, and population genetics. It also discusses the limits of deductive reasoning in the development of science.

In antiquity, the study of geometry was often perceived as identical to the study of nature, but the axioms of Euclidean geometry were gradually supplemented by the axioms of classical physics: conservation of mass, conservation of momentum, and conservation of energy. The significance of geometry as an organizing principle in nature was briefly subordinated by the discovery of relativity theory but restored in the 20th century by Emmy Noether's work on the relationships between conservation laws and symmetries. The book emphasizes the evolution of classical physics because classical insights remain the most important insights in many branches of science and engineering. The text also includes information on the relationship between the laws of classical physics and more recent discoveries that conflict with the classical model of nature. The main body of the text concludes with a section on the ways that probabilistic thought has sometimes supplanted older ideas about determinism. An interview with Dr. Renate Hagedorn about her work at the European Centre for Medium-Range Weather Forecasts (ECMWF), a leading center for research into meteorology and a place where many of the concepts described in this book are regularly put to the test, follows.

Of all mathematical disciplines, algebra has changed the most. While earlier generations of geometers would recognize—if not immediately understand—much of modern geometry as an extension of the subject that they had studied, it is doubtful that earlier generations of algebraists would recognize most of modern algebra as in any way related to the subject to which they devoted their time. *Algebra* details the regular revolutions in thought that have occurred in one of the most useful and vital areas of contemporary mathematics: Ancient proto-algebras, the concepts of algebra that originated in the Indian subcontinent and in the Middle East, the "reduction" of geometry to algebra begun by René Descartes, the abstract algebras that grew out of the work of Évariste Galois, the work of George Boole and some of the applications of his algebra, the theory of matrices, and the work of Emmy Noether are all described. Illustrative examples are also included. The book concludes with an interview with Dr. Bonita Saunders of the National Institute of Standards and Technology about her work on the Digital Library of Mathematical Functions, a project that mixes mathematics and science, computers and aesthetics.

New to the History of Mathematics set is *Beyond Geometry*, a volume that is devoted to set-theoretic topology. Modern mathematics is often divided into three broad disciplines: analysis, algebra, and topology. Of these three, topology is the least known to the general public. So removed from daily experience is topology that even its subject matter is difficult to describe in a few sentences, but over the course of its roughly 100-year history, topology has become central to much of analysis as well as an important area of inquiry in its own right.

The term *topology* is applied to two very different disciplines: set-theoretic topology (also known as general topology and point-set topology), and the very different discipline of algebraic topology. For two reasons, this volume deals almost exclusively with the former. First, set-theoretic topology evolved along lines that were, in a sense, classical, and so its goals and techniques, when viewed from a certain perspective, more closely resemble those of subjects that most readers have already studied or will soon encounter. Second, some of the results of set-theoretic topology are incorporated into elementary calculus courses. Neither of these statements is true for algebraic topology, which, while a very important branch of mathematics, is based on ideas and techniques that few will encounter until the senior year of an undergraduate education in mathematics.

The first few chapters of *Beyond Geometry* provide background information needed to put the basic ideas and goals of set-theoretic topology into context. They enable the reader to better appreciate the work of the pioneers in this field. The discoveries of Bolzano, Cantor, Dedekind, and Peano are described in some detail because they provided both the motivation and foundation for much early topological research. Special attention is also given to the foundational work of Felix Hausdorff.

Set-theoretic topology has also been associated with nationalism and unusual educational philosophies. The emergence of Warsaw, Poland, as a center for topological research prior to World War II was motivated, in part, by feelings of nationalism among Polish mathematicians, and the topologist R. L. Moore at the University of Texas produced many important topologists while employing a radical approach to education that remains controversial to this day. Japan was also a prominent center of topological research, and so it remains. The main body of the text concludes with some applications of topology, especially dimension theory, and topology as the foundation for the field of analysis. This volume contains an interview with Professor Scott Williams, an insightful thinker and pioneering topologist, on the nature of topological research and topology's place within mathematics.

The five revised editions contain a more comprehensive chronology, valid for all six volumes, an updated section of further resources, and many new color photos and line drawings. The visuals are an important part of each volume, as they enhance the narrative and illustrate a number of important (and very visual) ideas. The History of Mathematics should prove useful as a resource. It is also my hope that it will prove to be an enjoyable story to read—a tale of the evolution of some of humanity's most profound and most useful ideas.

ACKNOWLEDGMENTS

The author is deeply appreciative to Professor Karlis Podnieks for the generous way that he shared his time and insights. Also important to the production of this volume were the helpful suggestions from Frank K. Darmstadt, executive editor, and the expert photo research of Elizabeth Oakes.

Special thanks to Penelope Pillsbury and the staff of the Brownell Library, Essex Junction, Vermont, for their help with the difficult research questions that arose during the preparation of the book.

INTRODUCTION

Throughout history mathematicians and philosophers have speculated about the nature of numbers. Thoughtful individuals, often widely separated by geography and time, have made remarkable discoveries about numbers, and the study of numbers continues to generate new and interesting ideas as well as a numerical bestiary that includes natural numbers, the number zero, negative numbers, rational numbers, irrational numbers, algebraic numbers, transcendental numbers, complex numbers, floating point numbers, real numbers, transfinite numbers, and uncomputable numbers. Each type of number reveals something new about mathematics and perhaps the human imagination as well. One goal of this volume is to provide a history of these ideas; another goal is to give an accounting of our current understanding.

Historically progress in understanding numbers has often been slow. Sometimes the concepts associated with a new type of number were just difficult to develop quickly. Other times the concepts were fully understood by many, but it took generations for mathematicians and nonmathematicians alike to overcome the restrictions that they placed on their own concepts of what numbers are. Numbers provide a rich source of exotic ideas, philosophical and mathematical, but because many of us are so invested in the familiar—because many of us are so sure that we already know what numbers are—we have often resisted ideas about numbers that are new and unfamiliar.

The first section of *Numbers, Revised Edition* deals with numbers from the point of view of computation. Beginning with the earliest number concepts, it discusses the sophisticated computations of ancient Mesopotamian, Chinese, and Mayan mathematicians. It describes the origin and diffusion of Arabic numerals, and it concludes with a discussion of the way that the number system is represented within computers, including a new section that describes some of the IEEE (for definition, see page 60) standards for floating-point arithmetic.

However, numbers are about more than computation. The second part of this text discusses how the idea of number has broadened to include numbers that earlier generations of mathematicians rejected as being nonsensical. Beginning with the discovery of irrational numbers by the Pythagoreans, this history includes a description of the many contributions made by the classical mathematicians of the Indian subcontinent and the development of the concept of algebraic numbers by European mathematicians. This revised edition also contains a more complete description of the work of Richard Dedekind, especially the correspondence that he identified between numbers and points on a line, the correspondence that gave rise to today's real number line. In many ways, Dedekind's demonstration that the set of rational numbers taken together with the set of irrational numbers forms a continuum is the culmination of thousands of years of effort, but his accomplishment is not the end of this history. Even as Dedekind was formulating his ideas, research into numbers was taking a new tack as mathematicians and philosophers discovered (or created, depending on one's point of view) new concepts and new types of numbers.

The third section discusses the idea of infinity, a concept with which mathematicians and philosophers have long struggled. Sets of numbers provide a ready source of examples of infinite sets. In one of the most famous proofs of antiquity, Euclid showed that the set of prime numbers is infinite. (The proof is described in chapter 9.) This revised edition expands upon classical ideas by, for example, providing an exposition of the Greek ideas about the potential versus the completed infinite. Chapter 10 describes the first real progress in understanding the nature of infinite sets, which occurred during the 17th century with the work of Galileo. In retrospect, Galileo was far ahead of his time. It was not until almost two centuries later that Galileo's ideas were developed and refined by Bernhard Bolzano, who is also described in chapter 10. Chapter 11 describes the study of the infinite—and in particular transfinite numbers—begun by Georg Cantor. In many ways, Cantor's efforts formed the foundation on which modern mathematics is built, and his contributions are described in some detail. Logical difficulties associated with Cantor's formulation of set theory are also described. The text concludes with a description of the work of some of Cantor's most prominent successors: Bertrand Russell, David Hilbert, Kurt Gödel, Giuseppe Peano, and Alan Turing. Following Cantor's lead, Hilbert and Russell sought to axiomatize mathematics. Gödel and Turing demonstrated that there are limits beyond which the efforts to axiomatize mathematics cannot go. As a consequence of all this effort, still another type of number, the uncomputable number, was discovered during the latter half of the 20th century. The revised edition contains a more detailed discussion of the work of Gödel and Turing in order to prepare the reader for the discussion of uncomputable numbers, which is new to this edition. Finally, there is an interview with Karlis Podnieks, computer scientist and mathematical philosopher, about the nature of mathematics, its relation to the physical world, and the future of mathematical research.

The book also contains an updated glossary, chronology, and suggestions for further reading.

Numbers have attracted the attention of the practical and the philosophically minded. Few ideas have proven to be so rewarding from so many different points of view. The idea of number continues to grow in sophistication, and there is no end in sight. It is my hope that this book will be the beginning and that the reader will continue to learn about this simple-sounding idea from as many points of view as possible. Today's ideas about the concept of number are the product of 2,000 years of effort and represent one of humanity's most profound and useful set of insights.

PART ONE

NUMBERS FOR COMPUTATION

1

THE FIRST PROBLEMS

It would be interesting to know when people first began to develop a concept of number. There are no cases of a culture that developed a method of writing words without also—and simultaneously—developing a method of writing numbers. Just as early peoples knew how to speak before they began to write they must also have developed a concept of number before they developed a written language. As a consequence written records are of limited use in tracing the earliest ideas about number. The records do not go far enough back in time, and the prehistoric evidence is inconclusive. Scientists have unearthed stones from prehistoric sites that were carefully painted in decidedly nonrandom patterns. They have also found animal bones with numerous notches, carefully grouped and carefully carved, cut into the bones. Are these early attempts at recording numbers or are they just decorations? Without a "user's manual" it is difficult to know for sure.

It seems likely that early humans could compare quantities long before they could count them. There is often less technique involved in comparing two sets than in counting the elements of either. To understand how this applies to numbers, consider a common design for an early odometer, a device for measuring distances. Leonardo da Vinci designed one such device. It looked a lot like a wheelbarrow: The device had a single front wheel and the user moved it from behind by lifting the handles and pushing. The front wheel was attached through a series of gears to a circular device that revolved at the top of the odometer. This circular device was filled with pebbles. On the side of the odometer beneath the container of pebbles was a box. As the odometer was pushed forward the wheel rotated, and that rotation in turn caused the circular container of pebbles to turn. As the container of pebbles rotated it dropped a pebble—one pebble at a time—into the side box. Because the whole machine was geared to the wheel, each unit of distance traveled was represented by one pebble in the box. The farther the machine was pushed, the more pebbles dropped into the box. This measurement device established a *one-to-one correspondence* between the distance traveled and the number of pebbles in the box. There were exactly as many pebbles in the box as there were units of distance traveled by the odometer. Once the user knew the number of pebbles in the box, she or he also knew the distance traveled by the odometer. It is a simple idea, and it has far reaching implications for higher mathematics, but in itself Leonardo's concept of measurement tells the user little about the distance traveled by the odometer.

Although it may seem that Leonardo's odometer solves the problem of how to measure distances along a road, it does not. In fact it just shifts the problem from one of counting feet (or meters)

These painted pebbles were excavated from a cave in France. Approximately 1,400 painted pebbles have been discovered at this site. Painted about 11,000 years ago, their meaning is unknown. (© British Museum)

to one of counting pebbles. Leonardo's device establishes a correspondence between the number of units of distance traveled and the number of pebbles in the box, but at the end of the day, all the user really knows is that there are as many pebbles in the box as there are units of distance traveled. If you have ever tried to guess how many jelly beans are in a large full jar just by looking then you know that estimates can vary widely from one person to the next. It is rare for anyone to guess the number of beans in the jar correctly without emptying out the contents and counting the beans by hand.

Despite its shortcomings the simple idea of establishing a correspondence between one group of objects and another is valuable because it enables us to establish a clear definition of what a number is. In fact the idea that there must exist a one-to-one correspondence between any two sets of the same size—and that this correspondence is precisely the way we know that the two sets are the same size—lies at the heart of modern ideas about number. At some level it lies at the heart of all ideas about number, and so this abstract idea must have been formulated early in human history. People had to discover that five stones and five fingers have something in common. They had to discover that five stones and five fingers are both *instances* of the number 5. Notice that there is a clear distinction between examples or instances of the number 5 and the number 5 itself. There is only one number 5, but we are surrounded by many examples of the number 5. The *concept* of a number is a much deeper idea than an *example* of a number.

In the 20th century the eminent mathematician Bertrand Russell wrote an entire article, "The Definition of Number," in the hope of providing an unambiguous description of the concept of number. The fact that it took a highly creative mathematician to write such an article indicates that the subject is not as straightforward as it may at first seem. Russell wrote that the number 5, for example, *is* a set consisting of a collection of sets. Each set in the collection consists of five elements, so each set is an instance of the number 5. There is no set in the collection that is not an instance of the number 5, and all instances of the number 5 are included in the collection. In particular the set consisting of the letters *a*, *b*, *c*, *d*, and *e* belongs to the number 5. So does the set consisting of the states Alaska, Maine, South Carolina, New York, and Wisconsin. Many other five-element sets are included as well, of course. Describing a number as the collection of all instances of that number may seem awkward, even strange, but it has proved to be a very useful idea. Russell's definition, which we examine more closely later, arose as mathematicians worked to understand and classify sets that are infinite as well as finite. On some level his idea must have been shared by even the earliest of mathematicians who needed to recognize the equivalence of all sets with the same number of elements.

Having established a concept of number, the earliest mathematicians needed to find a way to record numbers. Part of what makes numbers so useful is that they enable us to classify sets of objects according to size. Ten bushels of grain is better than five. An army of 10,000 soldiers has the advantage over an army of 100 soldiers all other factors being equal. To *use* numbers to describe the world around them, these ancient peoples had to develop a method of counting. This was the second mathematical barrier to progress.

A nice historical example of a fairly modern counting method was used in Madagascar for the purpose of counting soldiers. Here is how it worked: The soldiers walked single file through a narrow passage or gate. As each soldier passed through the gate, a pebble was dropped into the 1s pile. Each pebble in this pile represented one soldier. With every 10th soldier the 1s pile was cleared away and an additional pebble was dropped in the 10s pile. The pattern was then resumed. Each pebble in the 10s pile represented 10 soldiers. When there were 10 pebbles in the 10s pile, that pile, too, was wiped clean and a single additional pebble added to the 100s pile. Notice that unlike Leonardo's odometer, in which every pebble represents one unit of distance, in the Malagasy scheme no pebble has a fixed value. The value a pebble acquires depends on the pile in which it is placed.

One advantage of the Malagasy method of counting is that "pebblewise" it is very economical. Whereas Leonardo's odometer, for example, would have required 123 pebbles to represent the number 123, the counters of Madagascar would have required only six pebbles: one pebble in the 100s pile, two in the 10s pile, and three in the 1s pile. In fact using the Madagascar idea, any counting number less than 100 can be represented by no more than 18 pebbles. (We never need more than nine pebbles in the 1s pile and nine pebbles in the 10s pile.) Similarly we need no more than three piles of nine pebbles to represent any counting number less than 1,000—nine in the 1s, nine in the 10s, and nine in the 100s—for a total of 27 pebbles. The larger the number, the more impressive the Malagasy approach appears. For example, any counting number less than 1 million can be represented by no more than 54 pebbles. This method of counting made it easy to keep track of large quantities of everything from soldiers to grain. It was an important conceptual breakthrough.

Notice that in the Malagasy scheme for counting soldiers the number 10 plays a special role. Each pile—the 1s pile, the 10s, the 100s, and so forth—can be in only one of 10 "states," where the word *states* refers to the number of stones in the pile. The pile may be empty—the "zero state"—or it may contain up to nine stones. There are no other possibilities. We describe this situation by saying that the Malagasy scheme is a base 10 scheme. It is one of the simplest and purest representations of counting in base 10 that we can imagine, but it can easily be modified so that one can count in any other base. In base 5, for example, the number 5 plays a special role: As the first soldier walked by, the counter would place a single stone in the 1s pile. Again each pebble in this pile represents one soldier. When there are five stones in the pile, the 1s pile is wiped clean and a single stone placed in the 5s pile. Each stone in the 5s pile represents five soldiers. When there are five stones in the 5s pile that pile is wiped clean and a single stone is placed in the 25s pile ($25 = 5 \times 5$). Each stone in this pile represents 25 soldiers. The stones are the same, of course, but the *meaning* of the stone again depends on the pile in which it is placed. In base 5 the pattern continues into the 125s pile ($125 = 5 \times 5 \times 5$), and when the 125s pile has five stones in it, it, too, is wiped clean and a single stone is placed in the 625s pile ($625 = 5 \times 5 \times 5 \times 5$). If we wanted to count in base 20 the same sort of technique would be used. The difference is that in this case we would have a 1s pile, a 20s pile, a 400s pile—because 400 = 20 × 20—and beyond the 400s pile there would be an 8,000s pile—because 8,000 = 20 × 20 × 20—and so on.

Conceptually we do exactly the same thing when we count in our number system. We do not use stones; we use numerals, but the principle is the same. Because we count in base 10, our "piles" represent powers of 10: ones, tens, hundreds, and so on. We seldom, if ever, count in a base other than 10, but there is no particular advantage to counting in base 10, and in some circumstances it is distinctly disadvantageous to do so. There are many instances in which mathematicians have found it convenient for one reason or another to count in a base other than 10. The Mayans of Central America, for example, counted in base 20. The Mesopotamians, as we soon see, counted in base 60. The chart to follow is modeled on the stone piling technique just described to show how different numbers would look in the bases 5, 10, and 20.

It is interesting to note that some cultures never developed a systematic method of counting. These same cultures often developed complex mythologies and varied forms of artistic expression, but they required few if any numbers to express their thoughts. Some groups of aboriginal Australians are said to have lived almost without numbers until recent historical times. Early anthropologists indicated that some groups of aboriginal Australians did not distinguish numbers much beyond 6 or 7, and that counting proceeded along the lines of "1, 2, 3, 4, 5, 6, many." We have to be careful in interpreting these early reports. Cultural and linguistic differences made it difficult for the anthropologists to communicate effectively with the people whom they studied, and their observations cannot be repeated since aboriginal Australians live in a different culture now and have adapted accordingly. In any case there is little doubt that not every culture developed a systematic method of counting. Not every culture developed a method of counting that enabled the user to count into the thousands or the millions or even higher.

At different times in history, sometimes independently of each other, people in various widely separated cultures began to come to terms with the idea of number. They developed a concept of

Number	Base 10			Base 5				Base 20		
	100s	10s	1s	125s	25s	5s	1s	400s	20	1s
7			•••• / •••			•	••			•••• / •••
26		• / •	••• / •••	•			•		•	••• / •••
123	•	• / •	•• / •		•• / ••	•• / ••	•• / •		••• / •••	•• / •
500	••• / ••			•• / ••				•	••• / ••	

The chart shows how the numbers 7, 26, 123, and 500 can be represented in bases 10, 5, and 20.

© Infobase Learning

number, and they made the first steps toward expressing numbers in a base that suited their needs. This slow evolution of the most rudimentary concepts about numbers occupied the greater part of the prehistory of the human race. The situation began to change, however, as people lived together in cities and together strove for the common good. There was an increased need for record keeping and for a system that enabled the user to perform ever more sophisticated computations. Progress began to accelerate.

2

EARLY COUNTING SYSTEMS

Mesopotamian civilization was centered in the area of the Tigris and Euphrates Rivers in what is now the country of Iraq. The inhabitants of this region invented one of the oldest of all written languages, and 5,000 years ago they began to write about themselves and the world around them. The Mesopotamian was one of the first of the world's great civilizations.

For better and for worse, Mesopotamian civilization was built on an open, difficult-to-defend land. As a consequence Mesopotamian cities were occasionally razed by conquering armies and occasionally enriched by immigration. This went on for millennia. The Mesopotamians adapted. Despite the onslaught of invading armies and the legions of newcomers Mesopotamian culture persisted. Cuneiform, their method of writing, was in use for about 3,000 years. For well over 1,000 years they had the most advanced mathematics on the planet. They also had, as far as is known, the first written code of laws, the Hammurabi Code, which consisted of 282 laws that governed a wide variety of social and legal situations. These people fashioned one of the most creative and durable civilizations in history.

Eventually, however, Mesopotamian civilization fell behind other cultures. The last known cuneiform records—records that contained information about astronomy—were made in the first century C.E. In time the culture of Mesopotamia was in large measure forgotten. It was not entirely forgotten, of course. The Hanging Gardens of Babylon were described by ancient travelers as one of the Seven Wonders of the Ancient World. There

are a number of references
to Babylon, one of the major
Mesopotamian cities, in the
Bible, and some of their ruins
have always been visible to
the interested traveler, but
the extent of the accom-
plishments of the people of
Mesopotamia is a fairly recent
historical discovery. Unlike
those of the Egyptians, who
built monuments so massive
and durable that they are
impossible to overlook, and
so serve as a constant remind-
er of Egyptian civilization,
most traces of Mesopotamian

This ancient cuneiform tablet, called Plimpton 322, is a list of triplets of natural numbers satisfying the algebraic equation $x^2 + y^2 = z^2$. The Mesopotamian's motivation for compiling the list is unknown. (Plimpton 322, Rare Book and Manuscript Library, Columbia University)

architecture were wiped out or buried beneath the sands. Most
of what we now know about Mesopotamia was unknown as late
as 150 years ago.

Mesopotamian civilization was rediscovered in the 19th century
as scholars began to decode the few examples of cuneiform writing
that were then known. It was slow, difficult work, but the more
they learned, the more interested they became. As archaeologists
went out into the field to uncover other Mesopotamian sites they
unearthed enormous numbers of clay tablets, which were covered
with cuneiform writing. Luckily for us, clay tablets proved to be an
extremely durable medium, and tens of thousands of tablets were
sometimes uncovered at single sites. The tablets address many
aspects of life in this ancient civilization. Archaeologists found letters sent home from students to parents. There were legal records.
There were stories and legends and poetry pressed into the clay.
There were tables of numbers, and astronomical computations,
and lots of math problems. It was an intimate portrait of one of
the first of the world's great civilizations. It is through these clay
tablets that we now know so much about this ancient culture and
its mathematics.

A MESOPOTAMIAN EDUCATION

We know quite a bit about Mesopotamian education because, among the many tablets detailing wars, business transactions, astronomical calculations, and government records, there are school books (or at least school tablets) and among these tablets there are a number of descriptions of a day in the life of a student. We know, for example, that the school day lasted from dawn to dusk. Students—practically all were boys—attended school for years. It was a difficult life. The following translation of a cuneiform text describes a day in the life of a schoolboy some 4,000 years ago.

> I recited my tablet, ate my lunch, prepared my (new) tablet, wrote it, finished it; then my model tablets were brought to me; and in the afternoon my exercise tablets were brought to me. When school was dismissed, I went home, entered the house, and found my father sitting there. I explained my exercise-tablets to my father, recited my tablet to him, and he was delighted.

(*Kramer, Samuel Noah*. The Sumerians, Their History, Culture, and Character. *Chicago: University of Chicago Press. Copyright 1963 by the University of Chicago. Used by permission)*

Not all stories are this positive, but except for the accounts of caning, which was the preferred method of disciplining students, many of the stories sound remarkably modern.

The Mesopotamian Number System

One of the main subjects taught in school was mathematics. The mathematics of the Mesopotamians was quite sophisticated. Becoming accustomed to it, however, takes a while because their number system was unique.

One of the characteristics that distinguishes Mesopotamian mathematics from all other systems of mathematics over the last 5,000 years is that the Mesopotamians used base 60 in their number system. The technical term for a base 60 system of notation is a *sexagesimal* system. Some remnants of this system survive today in our own number system. For example, we divide our minutes into 60 seconds and our hours into 60 minutes. (This gives 60 × 60 or 3,600 seconds in an hour.) To go back to the Malagasy method of counting soldiers with pebbles described in the previous chapter, counting in base 60 allows us to place up to 59 pebbles in the 1s pile. Each pebble in this pile represents an individual soldier. When the 60th pebble is placed in the 1s pile, it is wiped clean and a single pebble is placed in the 60s pile. Each pebble in the 60s pile represents 60 soldiers. When there are 60 pebbles in the 60s pile, the pile is wiped clean and a single pebble placed in the 3,600s pile (3,600 = 60 × 60). The choice of base 60 is unusual, and there has been a lot of speculation about why the Mesopotamians chose to count in this way. It turns out that there are some advantages to counting in base 60—advantages that we explore later in this chapter.

The Mesopotamians did more than count aloud. They also developed a way to write these numbers. (Writing, of course, is a separate problem from counting.) Cuneiform, the written language of Mesopotamia, was produced by pressing the end of a stylus into wet clay. The illustration on page 15 shows some Mesopotamian number symbols.

To write numbers larger than 59 they developed a system of *positional numeration*, which enabled them to "recycle" the first 59 symbols into the next position. The symbols remain the same but the *meaning* of the symbols changes, depending upon the column in which they appear. The concept is simply a written version of the counting method used in Madagascar. When counting with stones, the number that the pebble represents depends on the pile in which the pebble is placed. Similarly in positional numeration the meaning of the symbol depends on the column in which it is written. As with our system of positional numeration, the Mesopotamian system allows the user to "build up" expressions for very large numbers without expanding the number of symbols required to make these numbers. The Mesopotamians used 59 symbols to write numbers in their system; we use 10 symbols. Not every culture that developed a method of writing numbers also developed a method of positional numeration. It is to their credit that the Mesopotamians were able to do both.

The development of a written system of positional numeration by the Mesopotamians was a tremendous accomplishment. Their method is almost modern.

The difference in bases is not the fundamental difference between our system and that of the Mesopotamians, however. The essential difference between the two systems is revealed by the fact the we use 10 symbols to express our numbers in base 10, whereas for a long time they used 59 symbols to express their numbers in base 60. We use exactly as many symbols as the base in which we count. They used one fewer symbol than the base in which they counted. In hindsight it is easy to see that they were one symbol short. They were missing the symbol for zero.

The need for a symbol to represent nothing is not immediately clear to most people. The Mesopotamians were no exception. Initially they seemed to give no thought at all to the matter. Instead they relied on the context in which the symbols appeared to convey a sense of its meaning. To use the Madagascar example again: The Mesopotamians ignored the empty pile and allowed the reader to guess which pile represented which power of 60. For example, two 1s in adjacent columns might represent one group of 60 plus 1 ($1 \times 60 + 1$). We would write this number as 61. Or the same symbols might represent one group of 3,600 and one group of 60 ($1 \times 3,600 + 1 \times 60$)—the number that we would write as 3,660. (The 1s column was empty, but since they did not use 0 as a placeholder, there was nothing to mark its presence; in this situation the 1s column was absent.) Or two adjacent 1s might represent one group of 3,600 plus 1 ($1 \times 3,600 + 1$)—the number that we would write as 3,601. Presumably the difference between the number 3,660 and the number 61 is so large as to be clear to the reader from the context in which the symbols appear. The difference between 3,660 and 3,601 is not so apparent. Sometimes the Mesopotamians inserted a blank space between columns of symbols. This allowed them to distinguish between the numbers 3,601 and 3,660, for example. (Notice that 3,660 and 61 are still indistinguishable.) Leaving a space is clearly an improvement, but it is far from ideal since there is still a lot of ambiguity left in their notation.

Y	1	YY	2	YYY	3	YYY (symbol)	4
(symbol)	5	(symbol)	6	(symbol)	7	(symbol)	8
(symbol)	9	◀	10	◀Y	11	◀YY	12
◀YYY	13	◀(symbol)	14	◀(symbol)	15	◀(symbol)	16
◀(symbol)	17	◀(symbol)	18	◀(symbol)	19	◀◀	20
◀◀◀	30	(symbol)	40	(symbol)	50	Y	60

The chart shows how some of the numbers between 1 and 60 are represented in cuneiform. Notice the cuneiform addition problem. The Mesopotamian system does not include specialized mathematical symbols to represent addition and equality.

By the time the Mesopotamian region was conquered by Alexander the Great in 331 B.C.E., the Mesopotamians had invented a symbol to represent *some* of the empty columns that arise when writing numbers. If the empty column occurred between two nonempty columns, the Mesopotamians used two diagonal arrows to indicate the presence of the empty column. With this symbol the Mesopotamians were very close to finding a symbol equivalent to our 0, but they did not quite make the conceptual jump to a system of complete positional numeration. As far as is known the Mesopotamians did not use their symbol for an empty column when the empty column was the column farthest to the right—their 1s column. As a consequence this notation allowed them to distinguish easily between 3,601 and 61 but, again, not between the number 1 and the number 60, since in their system of positional notation both are written with their symbol for 1.

Of all the number systems of antiquity the Mesopotamian system most closely resembles a completely positional notation.

The Mesopotamian system of notation was never true positional notation, however. Right to the end—for a period of about 3,000 years—they continued to depend on the context in which the symbol appeared to distinguish between certain numbers. The discovery of a symbol for 0—in fact the concept of 0—would have to wait centuries until Hindu mathematicians invented the symbol, the same symbol that we use today.

Advantages of Base 60 Notation

The Mesopotamians went quite a bit further than what we have described so far: They also developed a system for fractions that was similar to our decimal system, although they continued to use base 60 rather than base 10. In what we would call the 10ths column, which is the column immediately to the right of the decimal point, they had the 1/60 column. In the position that we call the 100ths column, they had the $1/60^2$ (= 1/3,600) column. They included as many columns as they thought were necessary to maintain the accuracy of their calculations. The column immediately to the right of the $1/60^2$ column was the $1/60^3$ column ($1/60^3$ = 1/216,000), and so on. It is in the writing of numbers smaller than 1 that we can get some insight into why the Mesopotamians may have preferred a sexagesimal system. A sexagesimal system is, in some ways, more convenient for calculation than is our base 10 system.

Before we examine the Mesopotamian method of writing fractions we introduce some convenient notation that all descriptions of Mesopotamian mathematics employ. We need to introduce a punctuation mark that will work as our decimal point does. Since the Mesopotamians used sexagesimal notation instead of decimal, we do not use a point. Instead we use a semicolon (;) to separate the 1s column from the 1/60 column. (We use a semicolon so that we do not confuse base 10 numbers with numbers in base 60.) Furthermore we use a space to separate every column to the right of the semicolon. For example, the number that we might write as 1 2/60 is written as the sexagesimal number 1;02. (We mark the 1/60 column with two decimal digits—hence the 02 instead of simply the number 2.)

To see the advantage of computing with sexagesimal notation we need only consider a common fraction such as 1 2/3. In our decimal notation the number 1 2/3 is written 1.66666. . . . The *decimal*

MESOPOTAMIAN MATHEMATICS HOMEWORK

This is a translation of a "problem text," a clay tablet version of a textbook used for instructional purposes. We include it here because the numbers that appear in the problem are written in base 60. (The notation used in the problem is described in the main body of this chapter.) The problem itself uses a well-known mathematical result often attributed to the Greek mathematician Pythagoras: the Pythagorean theorem. Pythagoras, however, was born more than 1,000 years after this problem was written.

This text consists of the problem—which ends with the question, What is the breadth?—and an explanation of the algorithm used to solve the problem. The answer is identified as 0;10, and it concludes with the phrase "The method," which was the standard closing for Mesopotamian mathematics problems.

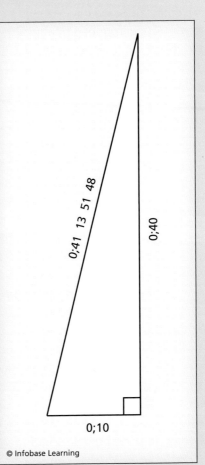

© Infobase Learning

Diagram of a problem described in the text. A cubit is approximately 18 inches (45 cm) long: The measurements are in sexagesimal notation.

The height is 0;40 (cubits), the diagonal 0;41 13 51 48. What is the breadth? You square 0;41 13 51 48. You will see 0;28 20. Square 0;40, the height. You will see 0;26 40. Take 0;26 40 from 0;28 20. You will see 0;01 40. What is the square root? The square root is 0;10, the breadth. The method.

(*Robson, Eleanor.* Mesopotamian Mathematics, 2100–1600 BC, Technical Constants in Bureaucracy and Education. *Oxford, United Kingdom. 1999. Reprinted by permission of Oxford University Press*) expression does not terminate; the 6s go on forever. The point at which you stop writing 6s and "roundoff" is the step in which you introduce an inaccuracy. But 2/3 can also be written as 40/60. So in base 60 the number 1 2/3 is written as 1;40. In sexagesimal notation the 2/3 is much easier to write and there is no round-off error. Here is one more example: In base 10 the number 2 1/12 is written as 2.083333. . . . The 3s carry on forever. But because 1/12 can also be written as 5/60, in base 60 the number 2 1/12 is written as 2;05. The sexagesimal expression is much simpler than the decimal expression of the same number and there is no round-off error.

It turns out that many common fractions are written much more simply in base 60—once you get used to it!—than they are in our own decimal system. This may explain why the Mesopotamians chose the sexagesimal system. It made their computations easier to perform accurately. We should not forget, however, that the lack of a 0 meant that the system that they employed for writing numbers with fractional parts had the same ambiguities that are to be found in their system for writing counting numbers.

The translation of ancient Mesopotamian texts and the search for their mathematical and cultural meanings remain active areas of scholarship. Part of the difficulty in understanding cuneiform texts is that the Mesopotamians had a very different approach to mathematics and a different understanding of the role of mathematics in society from ours. Scholars are still trying to sort out the differences. We know that they were interested in a simple kind of analytic geometry, in which geometric ideas were expressed in a sort of protoalgebra. Whatever we call it, we know that they put their knowledge to work when they divided estates, predicted astronomical phenomena, and calculated budgets for construction projects.

These applications in themselves do not explain the Mesopotamians' intense interest in mathematics. They computed extremely accurate approximations to certain numbers—the square root of 2 is a well-known example. The approximation, which is carried out to the 1,000,000th decimal place, is much more accurate than is required for practical work. Why did they bother? Did they just enjoy a mathematical challenge, or did the work have a deeper meaning to them? These are questions that are open for debate.

Much current scholarship centers less on what the texts say than on what the texts tell us about the role of mathematics in this early civilization.

The Egyptian Number System

Less information is available about Egyptian mathematics than about Mesopotamian mathematics, but one point is clear: Egyptian mathematics was not as advanced as Mesopotamian

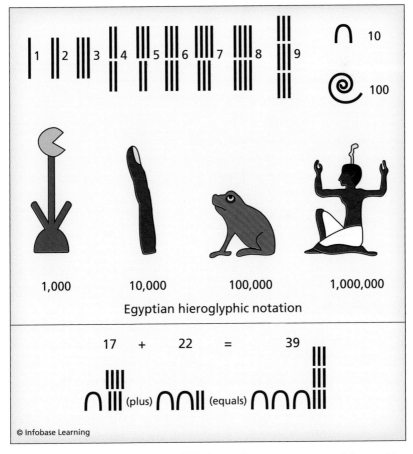

Egyptian hieroglyphic notation. Whole numbers are represented by combining the appropriate multiple of each symbol. Notice the addition problem in hieroglyphics. The Egyptians had no specialized mathematical symbols to represent addition and equality.

A PROBLEM FROM THE AHMES PAPYRUS

The Ahmes papyrus contains many problems that involve the equal distribution of bread and beer. Problems involving the distribution of bread and beer are not the only types of problems that appear in the Ahmes papyrus, however. There are, for example, geometry problems and problems that are simply computations without any additional motivation. Some of these problems would be difficult for many well-educated laypersons to solve today, but not all of them would. Today some would be considered trivial. At the time the Ahmes papyrus was written, however, the work involved in obtaining a solution to even a simple problem could be extensive because of the primitive notation and primitive methods of computation that the Egyptians used. Their math was primitive when compared with ours, of course, but it was also primitive compared with that of their contemporaries the Mesopotamians. It is clear from the problem included here that the system of notation used by the ancient Egyptians could make solving even simple problems difficult. The following bread problem is an example of an easy problem with a laborious solution. It is the fifth problem in the Ahmes papyrus, one of the simplest. The method of solution is described in the paragraph following the translation. (It is standard to identify those lines that are used in the computation of the final answer with a back slash.)

Problem 5

Divide 8 loaves (of bread) among 10 men.

Each man receives 2/3 1/10 1/30.
Proof. Multiply 2/3 1/10 1/30 by 10; the result is 8.

mathematics. One reason that less information is available about Egyptian mathematics is that the Egyptians wrote on papyrus whereas the Mesopotamians wrote on clay. Clay lasts longer. Much of what we do know about Egyptian mathematics is taken from two scrolls. The more famous of the two is called by one of two names, the Rhind papyrus or the Ahmes papyrus. Alexander Henry Rhind purchased the largest part of the papyrus while on a trip to Egypt in the 19th century. (Other parts were later located in the collection of the New-York Historical Society.) Ahmes was

Do it thus:	1	2/3	1/10	1/30
	\2	1 1/2	1/10	
	4	3 1/5		
	\8	6 1/3	1/15	

Total 8 loaves, which is correct.

(*Buffum Chase, Arnold.* The Rhind Mathematical Papyrus: Free Translation and Commentary with Selected Photographs, Transcriptions, Transliterations and Literal Translations. *Reston, Va.: National Council of Teachers of Mathematics, 1979.*)

Each entry in the right column represents what we would think of as a single number. (Keep in mind that the Egyptians used only the fraction 2/3 and unit fractions. So instead of writing 8/10 in the top row, Ahmes writes this number as 2/3 1/10 1/30, because 2/3 + 1/10 + 1/30 = 8/10. Each entry in the right column is a fraction or mixed number. To express it in a form that we would easily recognize just add all the fractions together.) Each entry in the right column is 8/10 the size of the corresponding number in the left column. Each of the entries in each row is twice as big as the corresponding entry in the row immediately above it.

For the Egyptians multiplication meant multiplication by 2. To multiply any two whole numbers, they just performed a series of doublings and kept track of each step. They solved the problem by adding some of the intermediate steps—here marked with a back slash (\)—to get their final answer. Here is his method:

To get 10 men, he adds 2 and 8 in the left column. Next he adds the corresponding numbers in the right column—that is 1 1/2 1/10 plus 6 1/3 1/15—to get 8, for the 8 loaves of bread. This proves that the first line is the correct answer: 1 man should receive 2/3 1/10 1/30 or 8/10 of a loaf of bread. It is a lot of work to solve such an easy problem.

the scribe who originally produced the text, which was written about 3,700 years ago. We refer to the text as the Ahmes papyrus, since, of the two, Ahmes did more work than Rhind. The other important text is called the Moscow papyrus. Less has been written about the Moscow papyrus, in part because it is a smaller work. The Ahmes papyrus was originally about 18 feet (5.5 m) long, and 13 inches (33 cm) wide. The Moscow papyrus is also about 18 feet (5.5 m) long, but it is only about 3 inches (8 cm) wide. Both scrolls are collections of mathematics problems.

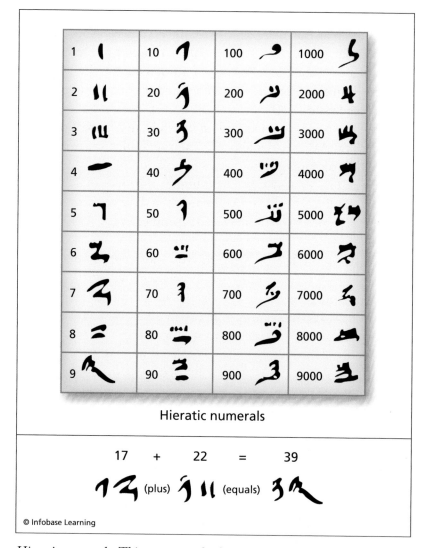

Hieratic numerals

17 + 22 = 39

© Infobase Learning

Hieratic numerals. This system used a large number of specialized number symbols. Its advantage is that the notation is more compact than hieroglyphic notation. Notice the addition problem in hieratic notation. The Egyptians had no specialized mathematical symbols to represent addition and equality.

Although the Egyptians counted in base 10 they had more than one way of writing numbers. The better-known form is hieroglyphic notation. In hieroglyphic notation there is no concept of positional numeration at all. Each power of 10 has its own symbol. For example, the number 1 is represented by a simple stroke, much as our own is. The number 10 is represented by a horseshoelike mark. The number 100 is represented by a curved line that is very similar to the top of a fiddlehead fern. To represent any counting number between 1 and 999 one simply wrote the required number of strokes, horseshoes, and fiddleheads. For example, 254 was written as two fiddleheads, five horseshoes, and four strokes. As a consequence in hieroglyphics the Egyptians could count as high as they had symbols for powers of 10, but no higher. If they needed to count higher they had to devise new symbols. Unlike the Mesopotamians, the Egyptians could not use hieroglyphics to write very large numbers without inventing new notation.

The writing in the Ahmes papyrus, however, is in hieratic script. Hieratic script was generally written in ink on papyrus, and it is a good deal more compact than hieroglyphics. That is its advantage. Its disadvantage is that there are separate symbols for 1, 2, 3, 4, 5, 6, 7, 8, and 9 as well as 10, 20, 30, . . . , 90; as well as 100, 200, 300, 400, . . . , 900; as well as 1,000, 2,000, 3,000, . . . , 9,000. To write the number 456 in hieratic script only three symbols are necessary, one symbol for 400, one symbol for 50, and one symbol for 6. To write the same number in hieroglyphics, 15 symbols are necessary: four symbols for 100 (four fiddleheads), five symbols for 10 (five horseshoes), and six symbols for the number 6 (six strokes). There was no concept of positional numeration in the hieratic system, either. If the order of the numerals was interchanged in hieratic notation the value of the resulting number remained the same.

For writing fractional parts of numbers the Egyptians had an unusual system of notation. This system influenced the development of the concept of number in nearby cultures, including that of the Greeks. The Egyptians used only fractions of a very special form. We would write this type of fraction with a counting number in the denominator and a 1 in the numerator. We call fractions with a whole number in the denominator and a 1 in the numerator *unit fractions*. The ancient Egyptians wrote these fractions by writing the counting number that appeared in the denominator and then drawing a dot or an oval above it. Because they considered only unit fractions, writing the denominator (with a dot) was enough to identify the fraction. For example, 1/2, 1/5, and 1/96 are all unit fractions. By contrast the number 3/4 is not a unit fraction, because it has a 3 in the numerator. (For the Egyptians the only exception to this rule was the fraction 2/3, for which they had a separate symbol.) Because they used only unit fractions, they had to write any other fraction as a sum of unit fractions. For example, the number we write as 3/4, they wrote as 1/2, 1/4 (1/2 + 1/4 = 3/4). Another example: The number that we would write as 21/30 they might write as 1/3, 1/5, 1/6, or as 1/2, 1/5, 1/30. Either set of fractions adds up to the number we know as 21/30. (There is generally more than one way to write a fraction as a sum of unit fractions.) Notice, too, that they did not put addition signs between the fractions. They maintained this system of writing fractions, and they computed with this somewhat awkward system, for thousands of years. From our viewpoint it is difficult to see any advantages to this type of notation.

The Mayan Number System

The Mayan civilization was centered in the area around the Yucatán Peninsula in what is now Mexico. Less is known about Mayan culture than is known of Mesopotamian or Egyptian culture because fewer records survive. We know that the Mayan written language used hieroglyphics, and we know that they had libraries with thousands of books. Unfortunately after the Spanish conquered the Mayans during the first part of the 16th century they set out to burn all Mayan books. They nearly succeeded. Today there are four more-or-less complete books left, the Dresden Codex, the Perez Codex, the Codex Tro-Cortesianus, and the Grolier Codex. None of these codices is in very good condition.

Mayan books were not constructed as the books that we know are. They are long sheets of paperlike material that are folded back and forth as a fan is. By flipping the folds the reader moves from page to page. The text with the most mathematics on it, the Dresden Codex, was damaged during the bombing of Dresden, Germany, late in World War II.

We can obtain a little more firsthand information about the Mayans from Spanish accounts written during and just after the conquest, although these accounts are sometimes more sensational than informative. In any case the Classic period, the time of greatest Mayan cultural innovation, ended centuries before the arrival of the Spanish. The Classic period lasted from about c.e. 300 to c.e. 900, when the Mayans, during a long period of extreme drought, abandoned many of their largest cities.

Despite these difficulties the Mayan number system is well worth investigation because it is unique and in certain aspects it is very sophisticated. Much of the Mayans' motivation for developing a sophisticated system of mathematics was their interest in astronomy and calendars. They worked hard to predict astronomical phenomena such as eclipses and to maintain an accurate calendar. The Mayan calendar was quite complicated and was at least as accurate as any calendar in the world at the time it was created.

The Mayan system of numeration is often described as base 20. This is not quite correct. The Mayans maintained two systems of numeration. One had religious significance and used a combination of multiples of 20 and multiples of 360. Although this system was important to the Mayans, Spanish accounts make clear that there was a second system of numeration in common usage. This "common system" *was* a base 20 system. This is the system on which we concentrate. Referring, again, to the story in the first chapter about the Malagasy system of counting, the Mayans would have placed up to 19 pebbles in the 1s pile. When the 20th pebble was to be added, they would have wiped the 1s pile clean and added a single pebble to the 20s pile. One stone in the first pile would represent a single unit, one stone in the second pile would represent 20 units, a single stone in the third pile would represent 400 units—20 × 20 = 400—and one stone in the fourth pile would represent a group of 8,000 (8,000 = 20 × 20 × 20). Just as our system is ordered around powers of 10, the common Mayan system

A page from the Dresden Codex, one of the few Mayan books to survive to modern times. Notice the numbers in each of the panels. (Center for the Advancement of Mesoamerican Studies, Inc.)

of numeration was organized around powers of 20. A base 20 system of numeration is a *vigesimal* system.

The Mayans expressed their numbers in their secondary or common system via true positional notation. In addition to their symbols for the digits 1 through 19 they had a symbol for 0. Theirs was one of the very few cultures in history that could see far enough into mathematics to recognize the need for, and the utility of, a symbol for nothing. The symbols that the Mayans used to represent their numbers are shown in the illustration. The symbol for 0 is represented by an oblong figure representing a shell. Each of the other 19 symbols in their system was represented by a combination of dots and dashes. Each dash represents 5. Each dot represents 1. The symbols were arranged vertically so that addition was a simple matter of merging the symbols in the corresponding positions and then, if necessary, "carrying" a digit. The Mayans apparently were quite skilled at this procedure and could perform extensive arithmetic manipulations with a few handfuls of beans. It was this facility with numbers that so impressed the Spanish colonists (see the illustration). This is a very convenient system for whole number arithmetic.

The Mayans spent a great deal of time developing their mathematics. Their astronomical calculations were surprisingly accurate. For example they calculated that 149 lunar months lasted 4,400 days. That is how *they* expressed their estimate of a lunar month. If we divide 4,400 days by 149 lunar months we get a more modern expression for the length of a lunar month. The difference

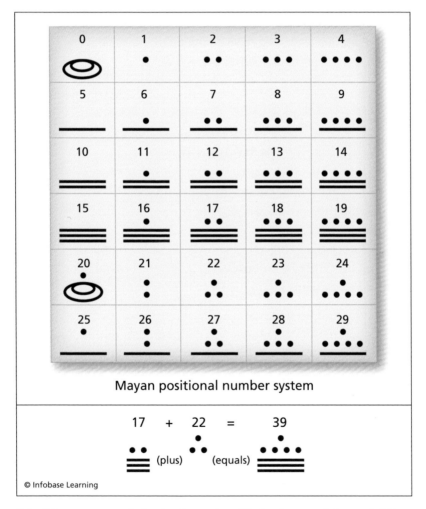

Mayan positional number system

© Infobase Learning

The Mayan system of vigesimal notation. Notice the use of the symbol for 0 in the number 20. This method was well suited for quick whole number calculations.

between the Mayan estimate for a lunar month and the actual value is about one minute, an extraordinary accomplishment, especially when one considers that they had no modern measurement devices and their mathematics *did not use fractions.* There is no evidence that the Mayans incorporated any sort of fractional notation in their vigesimal system. Since they had already solved all of the hard problems with respect to developing a complete system of positional notation it is difficult to see why they did not go further and develop a system that included vigesimal fractions. Apparently, however, what they invented was adequate for the computations that they wanted to perform.

The Chinese Number System

Chinese mathematics may be as ancient as that of Mesopotamia. Early records are scarce. We know that a system of writing was in use during the 17th century B.C.E. and that this early writing system had some similarities to contemporary Chinese writing. Presumably they produced a number of mathematics texts at this time, but none survives. In 221 B.C.E. China was united under the rule of Qin Shi Huang (also known as Shih Huang-ti), the first emperor of China. He established a strong central government that undertook many large-scale construction projects. Under Qin Shi Huang's rule the Chinese built an extensive system of roads, fortified the Great Wall, and instituted a uniform currency system. He also ordered the destruction of all books. This is why our knowledge of Chinese mathematics largely begins after the end of the reign of Shi Huang.

We do know that two systems of numeration were in use in ancient China. Both used base 10. One system used a symbol for each number from 1 to 10 and then a separate symbol for each power of 10 thereafter. To write a number one simply wrote one of the numerals (from 1 to 9) followed by the power of 10 the number represented. For example the number 321 would be written with the numeral for 3 followed by the symbol for 100 followed by the numeral for 2 followed by the symbol for 10 and

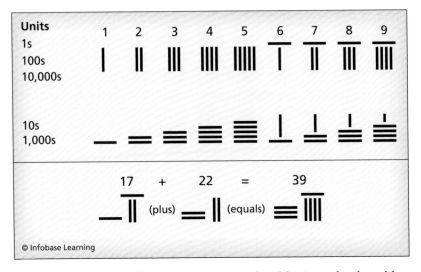

Chinese rod numerals. By alternating vertical and horizontal rods and by leaving a blank space for 0, the Chinese invented a system of notation that made arithmetical calculations quick and easy.

finally the numeral for 1. This is not quite positional notation but it is close.

The other system, called rod numerals, used only a set of straight lines to represent any counting number. The rod numerals are shown in the illustration. In this system there are separate symbols for the numbers 1 through 10 and for multiples of 10 up to 90: 20, 30, 40, . . ., 90. One can represent any counting number with these 18 symbols. Here is how: To represent the number 6,573, for example, write the symbol for 3, then the symbol for 70, followed by the symbol for 5, and finally the symbol for 60. (Alternating symbols in this way prevented ambiguity about where one symbol left off and the next began.) Where we would write a 0, the ancient Chinese left a blank space. This is almost a system of positional notation. Rod numerals were in use for a thousand years before Chinese mathematicians began using a symbol for 0. The symbol for 0 that they used is identical to the one we use today.

What is interesting about rod numerals is that they did not just resemble rods; in many cases the rod numerals were actually made with rods. Individuals who were required to perform arithmetic in the course of their job often carried a bag full of rods. Sometimes the rods were colored differently for positive and negative numbers. The rods were used as an aid to calculation. The numbers were constructed and then manipulated by adding and subtracting rods. A lot has been written about the speed and accuracy that these workers could attain with the help of their rods.

One of the earliest surviving Chinese mathematical texts is called *The Nine Chapters on the Mathematical Art* (also translated as *Arithmetic in Nine Sections*). The *Nine Chapters*, like the Ahmes papyrus, is a series of math problems on a variety of topics. There are problems in algebra and geometry as well as arithmetic. Not all the problems are correctly solved. Nevertheless, the problems are advanced for the time and so are many of the solutions. There are problems involving fractions as well as whole numbers, and some of the methods of solution that are employed in the text are similar to the methods that we use for doing arithmetic with fractions today. The book also contains fractions expressed in a kind of decimal notation as well as problems that represent a single fraction as a sum of unit fractions, in the manner of the ancient Egyptians. Some—though by no means all—of the problems in *The Nine Chapters on the Mathematical Art* have a very modern ring to them.

The effect of the *Nine Chapters* on subsequent generations of Chinese mathematicians was profound. Mathematicians wrote commentaries on the *Nine Chapters* for centuries after the book first appeared early in the Han dynasty. (The Han dynasty was established in 206 B.C.E., five years after the death of the first Chinese emperor, Qin Shi Huang.)

There are many other ancient cultures whose system of numeration we could examine. The Greeks, the Romans, the Etruscans, the Japanese, and others developed their own individual systems of numeration. The Greeks, for example, are famous for the

A PROBLEM FROM THE *NINE CHAPTERS*

The Nine Chapters on the Mathematical Art was described by a Chinese writer living in the third century C.E. as a text that survived the period of book burnings ordered by the emperor Qin Shi Huang and was revised and expanded upon by more than one author. The revised edition first appeared early in the Han dynasty, which lasted from 206 B.C.E. until C.E. 220. The problems in the *Nine Chapters* still pose a healthy challenge to many junior high and high school students. The analyses that follow the problems are not always easy for the modern reader to follow because of the lack of algebraic symbolism. They are just too wordy, but many of the problems (and a few of the analyses) have a modern ring to them; they could be from a contemporary textbook. Here are a very readable math problem and its accompanying solution from the Han dynasty. (Notice the use of the Pythagorean theorem in the solution of the problem.)

> Under a tree 20 feet high and three feet in circumference, there grows an arrowroot vine, which winds seven times (around) the stem of the tree and just reaches the top. How long is the vine?
> Rule. Take 7 × 3 for the second side (of a right triangle) and the tree's height for the first side. Then the hypotenuse is the length of the vine.

(*Mikami, Yoshio.* The Development of Mathematics in China and Japan. *New York: Chelsea Publishing, 1913*)

relative sophistication of their geometry, but their system of numeration was no more or less advanced than their neighbors'. The cultures that we have considered so far give a good sense of the mathematical diversity that was present many centuries ago. Each culture developed a unique system of numeration. They did not just differ in the way they wrote numbers or even in the bases that they used to express their ideas about numbers. There was also variation in their concept of what a number is. The Mesopotamians, for example, developed a very precise method of expressing arbitrary fractions. The Egyptians considered only unit fractions and the Mayans used no fractions at all. Each system had qualities that, in hindsight, facilitated or inhibited calculation, but each system, apparently, satisfied the needs of the users, whether they used the system to compute taxes or the dates of lunar eclipses.

3

OUR PLACE VALUE
NUMBER SYSTEM

The cultures of ancient China, Egypt, and Mesopotamia relied on experts to perform even simple calculations. They had no choice. Their systems of notation were so awkward that only the most practiced of mathematicians could perform arithmetic manipulations of moderate difficulty. (See, for example, the problem of dividing eight loaves of bread among 10 men in the Ahmes papyrus.) By contrast many of these previously difficult computational problems are now easy enough to be solved by schoolchildren.

The two great differences between arithmetic now and arithmetic in antiquity lie in the introduction of 0 and the truly positional place value notation that the concept of 0 made possible. The "invention of zero," although it seems almost too simple to matter, was a tremendous mathematical innovation. Many problems that were previously beyond reach became easily solvable. The frontiers of computation were greatly extended, and once a completely decimal system was adopted, computational techniques and science in general surged forward.

There is not much diversity left in the world with respect to systems of numeration used for hand calculations. Cultures all over the world now use base 10 to count, and it is a near-universal practice to employ 10 symbols—in the West those symbols are 0, 1, 2, . . ., 9—in a system of positional numeration. This is a testament to the superiority of the present system to what preceded it. Despite its importance to mathematics, not much is known about the way our current system of notation was established. Many conflicting reports exist. We do know that books and letters

Published in 1503, this woodcut shows two competing methods of compu-
tation. The figure on the left is using the "new" Arabic numerals. The
figure on the right is using a counting board, a device that is similar in
concept to an abacus. The figure in the center represents arithmetic.

describing the new system were written during the seventh century C.E., a time of interaction among many of the most mathematically advanced cultures of the age. Ideas about mathematics and

religion were exchanged between India and China to the east and between India and the emerging Islamic culture to the west, but the source of the individual contributions is often unclear. Indian mathematicians are usually given credit for devising our modern system of numeration, but it is not clear what they invented and what they borrowed. We have seen, for example, that centuries before the Indians, the Mesopotamians had an almost-complete system of positional notation, and that the Mesopotamians had a symbol that was used in a way that was somewhat similar, but not identical, to our own 0. The Mesopotamian system was base 60, however, not base 10. Others give the Egyptians credit for inventing a symbol for 0. Others give credit to the Greeks. One fact seems clear: The Hindu mathematicians were the first to join these ideas in one system of numeration.

It is customary to refer to mathematics that originated in the subcontinent of India as Hindu mathematics; unfortunately, there is not at present a more descriptive term. India is huge and culturally and religiously diverse. The Buddha was roughly a contemporary of Pythagoras (ca. 580 B.C.E.–ca. 500 B.C.E.). The history of the area is complicated and ancient. The mathematics that has emerged from this area is not uniform, and whether mathematicians in one part of the region were always aware of the work of mathematicians in another is not entirely clear. There are some common threads in the mathematics that originated in this area, however. One is a preference for algebra. Though the ancient mathematicians of present-day India almost certainly had some awareness of the geometry of the Greeks, they showed little interest in it. They were interested in algebra and trigonometry. (The sine function, the first of our six trigonometric functions, originated in India.)

The expositions of Hindu mathematics that have survived often read more like poetry than mathematics, and most of the math texts were, in fact, written in verse. This may have made them more fun to read, but it often caused the author to leave out the less poetic aspects of the subject matter, such as rational justification for certain formulas or techniques. Today no account of Hindu mathematics omits the assertion that it was in what is now India that our modern system of numeration was developed.

It is not clear when the mathematicians of this region began using what has become our system of numeration. We do know that in C.E. 662 a Syrian bishop named Severus Sebokt wrote a glowing description of the calculating system employed by Hindu mathematicians, a system that used *nine* symbols. This indicates two points: (1) They were using base 10 by C.E. 662, and (2) the account of Hindu arithmetic that had fallen into Sebokt's hands did not contain the symbol for 0. (Otherwise he would have asserted that they computed with 10 symbols.) Further we know that there are Indian documents from the late ninth century that contain the symbol for 0, so sometime between C.E. 662 and C.E. 900 the modern system of numeration was born. Hindu mathematicians did not extend their system of positional notation to decimals, preferring to use common fractions instead. (Common fractions are numbers that are expressed in the form *a/b* where *a* and *b* are whole numbers and *b* is not 0.)

It would be nice to report that the advantages of this new system of numeration were immediately recognized. This, however, was not always the case. Islamic culture was fairly quick to adapt. In the latter part of the eighth century a Hindu astronomical text, *Siddhānta*, was translated into Arabic. It was described in detail in the writings of the highly influential Islamic mathematician Mohammed ibn-Mūsā al-Khwārizmī (ca. 780–ca. 850). Soon the Hindu system of numeration was adopted throughout the Islamic world. In Europe, however, the process of change was slower.

There was quite a bit of resistance to the adoption of the Hindu ideas in Europe. The first known mathematical treatise to introduce the Hindu system of numeration to Europe and to argue for its adoption was written by Leonardo Fibonacci, also known as Leonardo of Pisa. His family was from Pisa, Italy, but they maintained a business in northern Africa. Leonardo was educated by a Moor—the Moors, who hailed largely from North Africa, conquered Spain in the eighth century—and as an adult Leonardo traveled throughout northern Africa and Syria. While there he was exposed to several systems of numeration. Some say that he learned of the Hindu system from al-Khwārizmī's book. Others assert that he learned it from a commentary on al-Khwārizmī's

EXPLAINING THE NEW SYSTEM

We are so accustomed to our system of numeration that it can be difficult to imagine teaching it as an innovation to adults familiar with another system of numeration. That is why it is so interesting to read the following excerpt of a 15th-century text called *A Treatise on the Numeration of Algorism,* an early European account of the so-called Hindu (or Arabic) system of notation. The author writes about the new system of notation with enthusiasm and a sense of wonder and seems to enjoy emphasizing how easy it is to write large numbers. (Notice that the author suggests writing each number from right to left.) But despite his evident enthusiasm, he slips back into Roman numerals. (In old English *nombre* means *number; ritht* means *right; sifre* means *cipher* or *zero;* the phrase *ten hundred thousand tymes* is how the author conveys the number *one million; composyt* means that the number is *composed* of different parts [the 1s part, 10s part, 100s part, etc.]; and the letters *y* and *v* are often used where we would use *i* and *u,* respectively.)

> Also ye schal vunderstonde that in nombrys composyt and in alle other nombrys that ben of diverse figurys ye schal begynne in the ritht syde and to rekene backwarde and so he schal be wryte as thus—1000. The sifre in the ritht side was first wryte and yit he tokeneth nothinge to the secunde no the thridde but thei maken that figure of 1 the more signyficatyf that comith after hem by as moche as he born oute of his first place where he schuld yf he stode ther tokene but one. And there he stondith nowe in the ferye place he tokeneth a thousand as by this rewle. In the first place he tokeneth but hymself. In the secunde place he tokeneth ten times hymself. In the thridde place he tokeneth an hundred tymes hymself. In the ferye he tokeneth a thousand tymes hymself. In the sexte place he tokeneth an hundred thousand tymes hymself. In the seventh place he tokeneth ten hundred thousand tymes hymself, &c. And ye schal vunderstand that this worde nombre is partyd into thre partes. Some is callyd nombred of digitys for alle ben digitys that ben withine ten as ix, viii, vii, v, iv, iii, ii, i. Articles ben alle thei that mow be devyded into nombrys of ten as xx, xxx, xl, and such other. Composittys be alle nombrys that ben componyd of a digit and of an articule as fourtene fyftene thrittene and suche other.

(Reprinted from The Earliest Arithmetics in English. *London: Oxford University Press, 1922, p. 70.)*

book written by another Islamic mathematician. In any case there seems little doubt that it was from Islamic mathematicians that Leonardo learned of the Hindu system of numeration, and it was through Leonardo of Pisa that the Hindu system of numeration was introduced into Europe. In his best-known book, *Liber Abaci,* he describes the Hindu method of numeration and advocates its use. Interestingly he uses the Hindu method only for whole numbers. When dealing with fractional parts of a whole number he sometimes retains the ancient Egyptian habit of confining himself to unit fractions. Other times he expresses himself in the sexagesimal fractions of the Mesopotamians. Leonardo Fibonacci had one foot in the future and one in the past.

The obvious question is, Why did Europeans take centuries to adopt the Hindu system of numeration? At least part of the answer lies in the fact that many Europeans recognized the Hindu system of numeration as what it was: a powerful tool for computation. There was, however, already a powerful computing device firmly entrenched in European society, the abacus. In fact those who supported the use of Hindu numeration were called algorists and those who opposed it were called abacists. The abacists were traditionalists who fiercely opposed the introduction of the new system. Both the Hindu system of notation and the abacus were perceived as being tools for quick and accurate computation. It may be difficult for us to see how positional numeration and the abacus could be perceived as mutually exclusive, but they were. The argument between the two groups went on for centuries.

The use of the Hindu system of numeration (for whole numbers) slowly became common practice throughout Europe, but until about 400 years ago fractions were still written as common fractions or in sexagesimal (base 60) notation. (If the continued use of base 60 sounds strange, it should not. We still measure time in a mixed system. We use base 60 for minutes and seconds, and we measure longitude and latitude in base 60 as well. The sexagesimal system of the Mesopotamians is one of the most enduring of all mathematical innovations.) By the late 16th century, however, there was great interest in developing more accurate, larger-scale methods of computation for both scientific and commercial purposes. The computations were carried out with the help of tables of numbers that were used in much the same way that we would use hand calculators. Bankers, surveyors, scientists, and others who worked with numbers depended on the tables to aid them in their calculations. The entries in the tables were usually recorded by using the Hindu system of numeration for the whole number and the sexagesimal system for the fractional part. It was exhausting to compile the tables and, given the mixing of decimal and sexagesimal systems, it must have been exhausting to use them.

It was during this time that we can begin to find decimal fractions appearing in the work of several mathematicians. No one person invented decimal fractions, but three people, François Viète, Simon Stevin, and John Napier, were particularly instrumental in promoting the acceptance of the new system. A complete base 10 system, including decimal fractions, made a huge difference in the science of computation.

The French mathematician François Viète (1540–1603) has priority. He was the first to present a reasoned argument against the continued use of the sexagesimal system for representing fractions and for the adoption of a decimal system in its place. He presented this argument in a supplement to his book *Canon Mathematicus* in 1579. In this work he tries to follow his own advice and uses several methods, more or less similar to our own, for representing decimal fractions. Where we would write 3.14, Viète sometimes wrote 3|14, using a vertical line where we would use a decimal point. Elsewhere he wrote 3$\underline{14}$ or even 3^{14}. He was searching for the best method of representing decimals, but his notation was in every case so similar to our own that we would have no problem following any of his notations.

The Dutch mathematician and engineer Simon Stevin (1548–1620) is generally given credit for doing the most to establish the decimal system of notation throughout Europe. Many writers give him all the credit. Here is why: Stevin wrote a small but very influential book describing the new system of notation. First published in 1585, *The Tenth* was not aimed at fellow mathematicians. Instead Stevin wrote the book—it was really more of a pamphlet—for anyone whose job required measurement and computation.

The introduction to the book is very explicit about the audience. Stevin states that surveyors, merchants, and people who work with tapestries, among others, all belong to his intended audience. In *The Tenth* he describes how one can use decimal fractions in addition, subtraction, multiplication, and division. He proves that his method of numeration, when used to do arithmetic, gives the correct answers, and, finally, he gives examples of how to use decimal notation when doing addition, subtraction, multiplication, and division. Though the book is short, it is very thorough. And because even very common fractions often yield "nonterminating" decimals—for example, 1/3 in decimal notation is 0.333333 . . .— Stevin even briefly discusses the effect of "round-off error," which occurs when we write a terminating fraction in place of a nonterminating fraction. For example, we might write 0.33 in place of 0.333 . . . and this introduces an error into the computation. Stevin assures the reader that one need only retain enough digits to get the desired accuracy. The ability to compute the thousandth part of an ounce in a business transaction, he says, is neither necessary nor desirable. He tells us that approximations can be more useful than perfection. His point of view is pragmatic and modern. (As we see in chapter 4, we follow Stevin's maxims whenever we use a calculator, whether we know it or not.)

The effect of the book was profound and immediate. It was quickly translated into several languages and readers took Stevin's advice to heart. After giving the cause of decimal fractions a big boost Stevin went further. He argued for the adoption of not just a system of decimal fractions but also decimal systems of weights and measures and money. All of this has happened. The metric system has been adopted by almost every country on Earth (the United States is a notable exception), and metric measurements are the standard for all scientists, including those working in the United States. Even though the United States still uses a nondecimal system of weights and measures, like that of most of the rest of the world, its monetary system is decimalized.

Interestingly, though, of the three mathematicians we mention here, Stevin's notation for decimal fractions was the most awkward and the least like our own. Unlike Viète's notation,

Stevin's notation is not nearly so easy for the modern reader to follow. In order to keep track of which is the 10ths column, the 100ths, and so on. Stevin placed the symbol ⓪, a 0 in a circle, above the 1s column of a number. He placed the symbol ① over the 10ths column. The symbol ② was placed over the 100ths column. The ③ was placed over the 1,000ths column, and so on. See the illustration for a comparison of some of these early notations.

The Scottish mathematician John Napier (1550–1617) was the third person who was

Viète: $2\underline{15}$ $\quad$ and $\quad$ $2\,|15$
$+3\underline{14}$ $\qquad\qquad$ $+3\,|14$
$5\underline{29}$ $\qquad\qquad\quad$ $5\,|29$

Stevin: ⓪ ① ②

$\quad$ 2 $\quad$ 1 $\quad$ 5
$+3$ $\quad$ 1 $\quad$ 4
$\qquad\qquad\qquad$
$\quad$ 5 $\quad$ 2 $\quad$ 9

Napier: 2.15
$+3.14$
$\qquad\quad$
5.29

© Infobase Learning

The equation 2.15 + 3.14 = 5.29 in the notations of Viète (two different methods), Stevin, and Napier

influential in the adoption of our current system of decimal notation. Napier was very interested in problems associated with computation. He recognized early in life that one of the great barriers to scientific progress in his time was the difficulty that scientists encountered whenever they performed calculations. He set out to find ways to facilitate computation. (Remember: Most scientists were still making their calculations with a system that used base 10 for integers and base 60 for fractions.) In a variety of ways Napier permanently changed the way calculations were performed. He discovered a way to facilitate simple multiplication through the use of a set of rods that had sequences of numbers on them. These are called Napier's rods or Napier's bones. The rods are sometimes used in math classes today to broaden students' appreciation of arithmetic.

More important for us, Napier found a way to facilitate very large-scale computations by inventing logarithms. Today we understand logarithms as functions. Logarithms are often studied in calculus courses, though in Napier's time calculus had not yet been invented, so his understanding of logarithms was somewhat different from ours. Suffice it to say that Napier communicated his results in the form of tables of numbers. His work was among the first to express numbers by using a decimal point between the 1s position and the tenths position.

Because Napier's discoveries were very useful, his text proved very popular. Napier's book was the ideal advertisement for his complete system of base 10 notation including decimal fractions. Napier preferred to use a decimal point instead of the notations that had been used by Viète and Stevin, and the success of his book led to the widespread adoption of the decimal point.

Although the decimal notation is not fully standardized—today some countries use a *decimal comma* where others use a decimal point—the system of numeration that we use today has passed to us largely unchanged from Napier's 1619 text on logarithms, *Mirifici Logarithmorum Canonis Constructio*.

One more innovation in numeration during this period occurred after the time of John Napier and has no relation to decimal fractions. The innovator was Gottfried Wilhelm Leibniz (1646–1716), the German mathematician who is best known as one of the two inventors of calculus. Leibniz had a deep interest in all matters philosophical, and this particular innovation was motivated by his interest in religion. He wanted a system of numeration that reflected his religious beliefs. He proposed a system of positional numeration that consisted of two symbols, 0 and 1. The symbols 0 and 1 would represent not just numbers but philosophical ideas. The 0 would represent the void; the 1 would represent God.

To illustrate how to count with just the two symbols we begin by (again) considering the system for counting soldiers that was first described in chapter 1. When the first soldier passes the counter a single pebble is placed in the 1s pile. When the second soldier passes the counter the 1s pile is wiped clean and a single pebble is placed in the 2s pile. As the third soldier passes the counter, a pebble is placed in the 1s pile. When the fourth soldier passes the counter the first two piles are wiped clean and a single pebble is placed in the 4s pile. The pattern can continue indefinitely. Every number, whether or not it is a whole number, can be expressed in base 2. From a computational viewpoint the advantage of base 2 is its simplicity. Its disadvantage is that even relatively small numbers can become unwieldy. The number that in base 10 is written as 100, for example, when expressed in base 2, has the form 1100100.

Leibniz's philosophical arguments for the adoption of base 2 numeration made little impression on his contemporaries. His idea gathered dust for more than two centuries, until the advent of digital computers. Computers are controlled by large collections of small circuits, each of which can be in one of two states, a *zero state* or a *one state*. Because most computations are now performed by computing machines rather than humans, we can see that after a slow start, Leibniz's idea about binary, or base 2, notation has triumphed.

With a true positional system of base 10 notation in hand, scientists and mathematicians were equipped with a tool of extraordinary power. Many computations that had previously required a great deal of time and effort were now almost trivial, and many computations that had once been too difficult to complete were now within reach. This development further accelerated progress in mathematics and science.

4

ANALYTICAL ENGINES

We have already remarked that arithmetic performed with positional numeration and decimal fractions is so simple that junior and senior high school students routinely perform the same kinds of computations that previously could be done only by experts. Actually our current system of decimal notation is even simpler than that. It is simple enough for a machine to use. Arithmetic operations in our current system of positional numeration are not just mechanical: They are mechanically simple. This is one of the great advantages of our system of numeration. In fact around 1619, at about the same time that John Napier was preparing his book *Mirifici Logarithmorum Canonis Constructio* for publication, a professor of astronomy, mathematics, and Hebrew at the University of Tübingen was designing a mechanical calculator. His name was Wilhelm Schickard (1592–1635).

Schickard's calculator could add and subtract automatically and, with some additional help from the user, could multiply and divide. Schickard was hoping to send his invention to Johannes Kepler, the author of Kepler's laws of planetary motion. Kepler's research required him to perform many calculations. Kepler and Schickard corresponded about the possibilities of such a device, but Schickard never completed his machine. The calculator was only partially completed when it was destroyed in a fire. Before Schickard could rebuild the device he died in an outbreak of bubonic plague. Because of these tragedies his work on the calculating machine had little influence on those who followed. His notes, which contain a detailed description of the machine, were rediscovered in the 20th century. His 20th-century successors

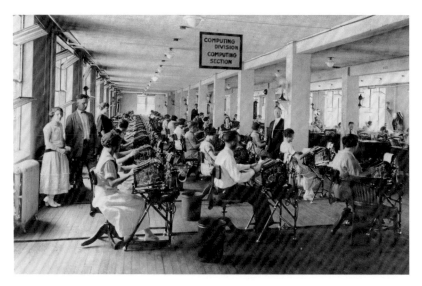

*Before the advent of the computer, large-scale computations had to be bro-
ken into small, human-scale subproblems. The solutions to the subproblems
were then combined to produce the final product.* (Library of Congress)

at the University of Tübingen used the notes to construct a model
of his invention, which is on display at the Deutches Museum in
Munich, Germany.

Shortly after Wilhelm Schickard invented his mechanical calculator another mathematician, Blaise Pascal (1623–62), built his
own mechanical calculator. Pascal was unaware of Schickard's
work. Instead Pascal's original intention was to help his father
with the many calculations he performed in his work as a government official. Eventually Pascal made a number of these
machines and sold some of them. A few still exist. The machines,
called Pascalines, were designed for addition and subtraction;
one could not use them for division and multiplication. By all
accounts Pascal's machines were of a less sophisticated design
than Schickard's. Their importance and their beauty lie in the
concept behind the calculations. Mechanically they consist of a
series of wheels with the numerals 0 through 9 on them. The
wheels are connected to each other in such a way that when one
wheel is rotated 360° the wheel immediately to the left rotates

1/10 of a revolution or 36°. This is the mechanical version of "carrying the 1." Here we see an example of our positional system of numeration represented not with 10 symbols but with wheels

CALCULATORS, COMPUTERS, AND THE HUMAN IMAGINATION

Schickard's, Pascal's and Leibniz's mechanical calculators share an important characteristic with even the most modern computers: The set of numbers that machines are designed to manipulate is fundamentally different from the set of numbers that we humans imagine.

Our number system, as we envision it, is infinite. To any number, however large, we can, in theory, always add a 1 and obtain an even larger number. And any positive number, however small, can always be multiplied by 1/2 to obtain a number that is still bigger than 0 but only half the size of the original. We can also add very large numbers and very small numbers together, and our answer will be different from both of the two numbers with which we begin. No person questions the truth of these obvious-looking statements, but no machine can implement them, either. No matter what number we enter into the Pascaline, if we repeatedly add a 1 to it, we eventually exhaust the capacity of the machine to add. This is also true of all the successors of the Pascaline.

This limitation did not make these mechanical calculators useless. As long as their use was restricted to the small collection of arithmetic problems that they were able to solve, the machines relieved the user of much of the tedium associated with routine calculation. This is precisely what they were designed to do. Still the limitations of the early calculators must have been evident to even the occasional user. Mechanically Pascal, Schickard, Leibiz, and others had found a way to represent part—but only part—of our number system through a system of gears and levers. There would, however, never be enough gears and levers to represent all of the numbers that they were able to imagine.

As mechanical calculators made way for much more powerful electronic computers, gears and levers gave way to transistors. But the limitations of these new machines, though less obvious to most users, are still present. The limitations still arise from the very fundamental difference between our infinite system of numbers and its finite representation inside the machine. In one form or another this difference between machines and the human imagination is maintained to the present day.

and gears. This is the advantage of our system of numeration. It is so mechanical that it can be represented by gears and levers or even electronic circuits!

The Pascaline could also add decimal fractions as well as whole numbers. Because we carry the 1 when adding decimal fractions in just the same way we carry the 1 when we add whole numbers, the Pascaline is just as capable of adding decimal fractions as it is whole numbers. The key is to enter each number in the Pascaline with the same number of digits to the right of the decimal point. For example, if one wanted to add the numbers 3.41 and 11.1, one would need to enter 3.41 and 11.10 into the Pascaline and then remember to interpret the answer as 14.51 instead of 1451. This is entirely equivalent to "lining up" the decimal points when we add a list of numbers by hand.

Pascal did not sell many of these calculators. There was not much demand. The manufacturing and assembly processes were crude, so the parts did not always fit together well. As a consequence they were not mechanically reliable. In addition to the mechanical problems, the machines were expensive. Nevertheless Pascal's was one of the first more or less successful efforts at mechanizing arithmetic processes.

For the next two and one-half centuries mathematicians and inventors worked at designing better versions of the Pascaline. Essentially all of these devices used the base 10 system of positional numeration represented as a series of interlocking gears. Inventors hoped that these devices would free people from the tedium of calculation.

One of the earliest and most notable improvements on Pascal's design was made by Gottfried Wilhelm Leibniz. Leibniz was one of the most versatile and creative mathematicians in history. We noted in chapter 3 that Leibniz developed the binary system of numeration and coinvented calculus. Leibniz acknowledged that the part of his machine that performed additions was identical to the Pascaline, but his invention could also multiply. It did this through a device now known as a Leibniz wheel. Mechanically Leibniz's calculator represented multiplication as a series of additions. The process of multiplying one number by another was achieved by turning the handle of a crank. This invention influenced generations of designers.

Charles Babbage and the Analytical Engine

The next breakthrough in computation occurred when the British mathematician and inventor Charles Babbage (1791–1871) began to think about what he called an analytical engine. Charles Babbage's analytical engine was the forerunner of the modern computer. In some ways he anticipated the design of modern computers, although the language that he used to describe his ideas definitely is that of another age. For example, he imagined that numbers would be kept in the "store," which, today, we call the memory, and that the processing would occur in a different part of the computer, which he called the "mill." What Babbage called the "mill" we call the central processing unit (CPU). He imagined that the computer would receive its instructions via punched cards, and in all of these ideas he generally anticipated how modern computers would be designed. (Of course modern computers do not use punch cards, but the first few generations of electronic digital computers could not work without them.) On the other hand he wanted to run his computer with a steam engine, which was the most modern power source of the era. (Steam-powered computers never caught on.) More importantly for us he thought that each number should be entered into the store and the mill with the same number of decimal digits.

It is important to understand Babbage's ideas about how numbers should be represented in computers, because these ideas guided the first modern attempts at building electronic computers. Babbage envisioned that his engine would store and manipulate numbers expressed in our system of base 10, positional numeration. Each digit would be stored in a place, or "cage," in the machine—one digit per cage. Because the mechanics of manipulating decimal fractions are exactly the same as those involved in manipulating whole numbers, Babbage said that if one wanted to work with decimals one could just imagine where the decimal point would be and then interpret the results accordingly. We call

A precursor to his never-finished analytical engine, Charles Babbage's difference engine was designed to perform computations associated with astronomy, navigation, and engineering. (Niagara College)

this method of representing numbers fixed-point notation because the decimal point remains in the same place throughout the computation. In other words if one number in the machine contained a 10ths column then *every number* entered into the machine should contain a 10ths column, even if the 10ths entry were a 0. Babbage did this so that all decimal points would be lined up in much the same way that we line up the decimal points whenever we do arithmetic by hand. As with the Pascaline, Babbage's analytical engine could only operate with the decimal points aligned. If every number were entered into the analytical engine with the same number of decimal digits, then the user was only required to remember where the decimal point was located initially, because the result was also expressed with the same number of decimal digits. The engine kept track of the rest of the arithmetic.

One difficulty with fixed-point arithmetic is that it severely limits the quantity of numbers with which the machine can work. For example, suppose we decide to store each number in four of what Babbage called cages. (Babbage proposed a larger number of cages, but the principle is exactly the same.) Each of the cages holds exactly one digit, so an analytical engine with four cages can manipulate only three-digit numbers. The difference between the number of cages and the number of digits arises because the first cage holds the sign of the number: The first cage indicates whether the number is positive or negative. The next three cages must hold the digits that most closely approximate the number that we have in mind. If we deal in whole numbers this means that we can write any whole number from –999 up to +999. If, however, we decide to use two decimal places for increased accuracy, then, *because every number must also be represented with those same two decimal places*, we can store only numbers from –9.99 to +9.99. The range of numbers that we can represent is 100 times smaller! Part of the difficulty arises from the fact that we are trying to represent our infinite number system in a computer with finite storage capacity, but another source of difficulty lies in the fixed-point system itself. With fixed-point arithmetic what we gain in accuracy we lose in the range of numbers available for computation. The solution is not immediately obvious. Mathematicians and engineers devoted a lot of time to finding the most efficient way of representing our number system in a computer before arriving at the system presently in wide use.

Babbage never completed his machine. It seems that whenever he worked on it, he found a better way to design it and so would begin again from the beginning. He continued to refine his theory long after others lost interest in his project. Babbage was far ahead

AN EARLY ELECTRONIC REPRESENTATION
OF OUR NUMBER SYSTEM

Early research papers on computer design show that although research-ers were making use of new electronic technologies, their conception of numbers was still old-fashioned. Initially they organized the circuits so that the computer could store numbers and perform arithmetic opera-tions in our base 10 system of positional numeration. They still favored Babbage's fixed-point arithmetic as well. Consider this excerpt from an early paper that explains one way of representing numbers and perform-ing additions and subtractions:

> There are assemblages of *(ten)* thermionic tubes with associ-ated circuits, having the following properties. One tube of the ring alone is in conducting condition. On receipt of a pulse this tube becomes unexcited, and the next in order becomes excited. The direction of succession around the ring, positive or negative, is determined by which one of a pair of associated electronic switches is energized. The ring thus counts pulses. A restoring circuit is provided, energizing of which restores the ring to the condition where the zero tube only is energized. . . . An activating circuit may also be associated with such a ring. When this is energized the ring steps automatically at a prede-termined rate and in a predetermined direction back to the zero position, and on each step emits a pulse into a receiving circuit.

(Bush, Vannevar. "Arithmetical Machine." Vannevar Bush Papers. Container 18, Folder: Caldwell, Samuel, 1939–1940. Library of Congress, Washington, D.C.)

This may sound familiar. You have encountered this idea earlier in our story. It is a vacuum tube version of the Pascaline! The "assemblage of (ten) thermionic tubes" is the electronic version of a gearwheel. The sentence, "On receipt of a pulse this tube becomes unexcited, and the next in order becomes excited," is the electronic version of turning the gearwheel 36 degrees or 1/10 of a revolution. Turning the Pacaline's gearwheel in one direction represented addition and turning it in the other direction represented subtraction. In the electronic version, a "pair of associated electronic switches" controls the direction in which the tubes are energized, and the "direction of succession around the ring" is used to represent inverse arithmetical operations. Nearly 300 years after the Pascaline was first created, early computer designs were still heavily influenced by its simple collection of interlocking gears!

of his time in imagining how a computer should work. A lot of the architecture of a modern computer that we take for granted was first imagined by Babbage, but he was wrong about how numbers should be stored. Babbage's system of fixed-point arithmetic would lead to substantial difficulties in even relatively simple calculations. The problem is that Babbage's system is incapable of representing the wide variety of numbers that arise in many applications.

The Representation of Numbers in Computers

Computer design did not attract much attention until reasonably reliable vacuum tubes were developed in the first decades of the 20th century. The vacuum tube was the predecessor to today's transistor. Both tubes and transistors are designed to control and modify the flow of electrical current. Vacuum tubes, however, were slower to respond to changes in the circuit. They also required considerably more energy to perform the same function, but at the time vacuum tube technology seemed extraordinarily fast and efficient. Much of the original motivation for the development of computers stemmed from their ability to speed progress in computationally intensive fields, especially ballistics calculations, and later radar, sonar, and nuclear weapons research. One of the barriers to progress in all of these fields was the slowness with which calculations could be completed.

Computer design was greatly accelerated by World War II (1939–45). The first working electronic computer was called the Electronic Numerical Integrator and Computer (ENIAC). It was finished in 1946, one year after the end of World War II. ENIAC was a huge machine that contained 18,000 vacuum tubes. It was a big step forward, but it continued to store and manipulate numbers in base 10 in a way that was somewhat similar to the way that Babbage proposed.

ENIAC and each of its immediate successors were individually designed and constructed. Each machine was very expensive and labor-intensive. Each represented a tremendous effort not just to build a computer but to overcome the limitations discovered in the previous machines. Almost as soon as ENIAC was built John

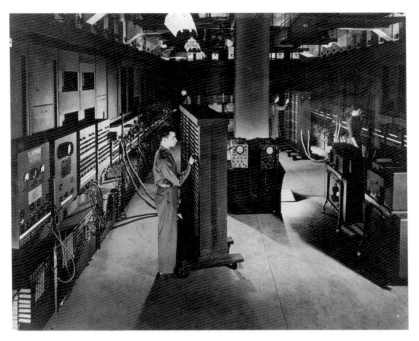

The Electronic Numerical Integrator and Computer (ENIAC) was the world's first digital computer. (John W. Mauchly Papers, Rare Book & Manuscript Library, University of Pennsylvania)

von Neumann (1903–57), a prolific mathematician who was very involved in the theory and design of these machines, saw that base 2 represented a considerable gain in simplicity over the base 10 system that ENIAC had used. Because the circuits could assume only one of two states—the circuit was either off or on—base 2, the discovery of Gottfried Wilhelm Leibniz centuries earlier, was a more natural choice than base 10 for the computer. It simplified the design of the circuits.

Changing the base in which numbers are stored does not, however, address the problems associated with fixed-point arithmetic, and there was some controversy about how this problem could be solved. There were two choices: Designers could continue to work with fixed-point arithmetic, or they could incorporate what is called floating-point representation into the design of the machine. Floating-point arithmetic is more flexible in that it can better store numbers of widely varying sizes. There were those who believed that fixed-point arithmetic was the more efficient of the two methods of representing numbers. But other devices already being constructed in other places used the floating-point form, and the success of the new idea quickly convinced everyone that what we now call floating-point representation was the better of the two alternatives.

Floating-Point Representation

One of the fundamental jobs a computer has is to store numbers. The number may be obtained from an input file, or it may be the result of a computation that the computer has just finished performing. Very often the computer does not store an exact copy of this number in its memory. Instead it stores an approximation. Much of the process of approximation is unrelated to the base in which the number is expressed. It relates to the limitations of the machine itself. Because a computer has only a finite amount of memory and there are infinitely many numbers, designers must be careful to ensure that the computer does not waste all of its memory storing a few or even just one very long number. We are usually unaware of how computers represent numbers because modern computers are so good at doing so that we do not usually notice their shortcomings. Of course many numbers can be stored in the computer's memory with perfect precision, but most cannot. The problem that concerns us arises when the computer stores a number that requires many digits to be expressed. For example 1/3 as a decimal fraction is 0.333 . . . The sequence of 3s does not terminate. Another example is a very large integer that has many nonzero digits. These kinds of numbers represent an additional challenge to computer designers because they *cannot* be stored with perfect accuracy. There will never be enough memory to do so.

The difference between the number that we want to store and the number that the computer actually places in memory is called the round-off error. Round-off error is unavoidable because the computer designer can allocate only a fixed amount of storage space for each number. Although there is great variety in the numbers to be stored, the amount of storage space is the same for every number. Insofar as is possible, this storage space must hold all the vital facts about the number it contains. These facts include

- The sign of the number (i.e., whether it is positive or negative)

- The digits that make up the number

- The size of the number (Remember that in a positional system of numeration it is the position of each digit that tells us its value. Representing the digits that constitute the number and the size of the number are distinct problems.)

When accomplishing these tasks, computers must also represent the number in a form convenient for storage. We can think of this stored form of the number as a kind of scientific notation. When we write a number in scientific notation we write it as a product of two numbers. The first number is greater than or equal to 1 and less than 10. The number, called the mantissa, gives us information about the digits that make up the original number. The second number is a power of 10. The power of 10 that appears gives us information about the size of the number. Consider, for example, the number 172, which, when written in scientific notation, is 1.72×10^2. The number 1.72 is called the mantissa. The superscript 2 is called the exponent. One way of understanding the exponent is that it is an indicator: The exponent indicates how many places we must move the decimal point in the mantissa to regain our original number. Alternatively we can think of the exponent as a measure of the size of the number: When the exponent is positive we move the decimal point to the right; when the exponent is negative we move the decimal point to the left. (Computers, as already noted, do not use base 10 to store numbers. We use base 10 here because most readers are more familiar with base 10 and because the principle is the same for any base.)

FLOATING-POINT ARITHMETIC
AND YOUR CALCULATOR

Here is a simple example of how computers solve certain problems in a way that is fundamentally different from ours. Consider the following two equations.

The author solved both equations with his "scientific" programmable calculator. These two equations differ only in the order in which the operations are performed. A quick check should convince the reader that the order of operations should not matter in this case. Both equations have the same answer. The answer should be 0.00001 for both equations, but in the first equation the calculator yielded the wrong answer of 0, and in the second equation the calculator gave the correct answer of 0.0001. The programmable calculator that

$$100,000 + 0.00001 - 100,000 = x$$

$$100,000 - 100,000 + 0.00001 = x$$

© Infobase Learning

Although these two equations have the same solution, most calculators produce distinct solutions because of the way that numbers are represented in machines.

We can get some feeling for the way this works in practice if we imagine the storage space allocated for a number to be a series of boxes, or, as Babbage would say, cages. In each box we place one fact about the number. Part of the storage space contains information about the sign of the number—specifically whether the number is positive or negative—and part of the storage space contains information about the digits that make up the mantissa. Finally, part of the storage space contains two pieces of information about the exponent: the sign of the exponent and the value of the exponent.

To see how this works, imagine storing a number in a computer. We can imagine that our "number storage compartment" is designed to hold 10 facts about our number. All the information that the machine retains about our number must fit inside those was used works on the same principles as a more powerful and more sophisticated computer.

Here is why the programmable calculator gave two different answers to what was essentially the same problem. The numbers were entered into the calculator from left to right. In the first case the computer added 100,000 and 0.00001 to obtain 100,000.00001. Notice how long the mantissa is in base 10: It is 11 digits long. When the calculator stored this number in memory, the actual mantissa, which in base 10 we write as 1.0000000001, was too long to fit into the space allocated for its storage. The calculator responded as it was programmed to do: It rounded off the mantissa to 1. So after the mantissa is rounded off, the result of the first operation is lost. It is as if it never occurred. As far as the calculator is concerned 100,000 + 0.00001 is exactly the same as 100,000. Next the calculator subtracts 100,000 from 100,000. The result is 0.

In the second equation we begin by subtracting 100,000 from 100,000. The result, which is stored in memory, is 0. Now we add 0.00001 to obtain the correct answer of 0.00001. To us the two equations are equivalent. To the author's calculator they are not! It is not hard to make up problems for which a more powerful computer will likewise yield the wrong answer. This is a fundamental problem that stems from the difference between our infinite number system and the finite number system used by all computers. The number system that we imagine is not the same system that is implemented on our machines.

10 places. There are no exceptions. If there is not enough room to fit all of the necessary facts then we have to omit enough facts to obtain a fit.

First, imagine an empty number storage compartment:

$$| \, | \, | \, | \, | \, | \, | \, | \, | \, |$$

Notice that there are 10 compartments, or cages, to hold all the facts about our number. For purposes of illustration we store the number 123,456,789. (In scientific notation we write 123,456,789 as 1.23456789×10^8.)

The first fact we store is the sign of the number. The next six places hold facts about the mantissa. The place immediately to the right of the mantissa holds the sign of the exponent, and finally we keep two places for information about the exponent. We store the number in four steps:

- We begin by placing a + sign in the first position because our number is positive. Now our compartment looks like this: |+| | | | | | | | | |.

- Now we place *as much of the mantissa as we can fit* into the next six slots. At this point our compartment looks like this: |+|1|2|3|4|5|7 | | | | |. (Notice that there is not room enough to fit the entire mantissa, so we round it off and use the first six digits. Notice, too, that because every mantissa is greater than or equal to 1 and less than 10, it is not necessary to store the decimal point because it is always in the same place—immediately after the first digit of the mantissa.)

- The exponent is a *positive* 8, so we place a + sign in the next box: |+|1|2|3|4|5|7 |+| | |.

- Finally, because the exponent is 8 we write a 08 in the remaining two empty boxes. Our number compartment looks like this: |+|1|2|3|4|5|7 |+|0|8|. (Notice that other numbers, for example, the numbers 123,457,000 and 123,456,955, are also rounded off to exactly the same number as that stored in our imaginary computer.)

One great advantage of floating-point notation is that it is adaptive. It can store numbers of widely varying sizes. It is also more useful for storing *significant* figures, and the more positions set aside to store the mantissa the less likely it is that important information will be lost. In fact the precision of the machine is often measured by the number of "cages" set aside to store information about the mantissa.

Floating-point notation has proved to be a much more flexible system than the fixed-point notation that was envisioned by Babbage and implemented in the Electronic Numerical Integrator and Calculator (ENIAC). Floating-point notation is not perfect (see the accompanying sidebar), but it is so far the best method available for representing our infinite set of numbers in a machine that can store only a finite amount of information. After its introduction in the 1950s it quickly displaced most fixed-point applications. Today desktop computers and supercomputers alike use a floating-point system to store and manipulate numbers.

Although most people continue to use the strictly positional, base 10 system of notation employed by Napier about four centuries earlier, this is not the system of notation most in use today. Because machines now perform the vast majority of all computations, floating-point notation—a system with which very few of us have any direct experience—is probably more characteristic of our time than the decimal system of notation pioneered by Napier, Stevin, and Viète.

More about Computers and Numbers

The set of all real numbers, the set that is first introduced to students in middle school, is something that had to be created (or discovered, depending on one's point of view). It is only one of many ways of thinking about numbers, and its structure reflects philosophical concerns as much as it reflects arithmetic ones. The creation of the set of real numbers was, in part, a response to the desire to have a set of numbers that forms an unbounded continuum. (To say that the set of real numbers forms a continuum means that there are as many real numbers as points on a line. See chapter 8 for a discussion of this very important idea. The term *unbounded* means that no matter how large a number we imagine, the set of real numbers contains a number that is still larger.) Mathematicians have found the current definition of real number to be convenient because it facilitates mathematical research, but it was not created with computation in mind. It is, in fact, a definition of number that is not especially suited for computation.

Computers use a number system with properties that are very different from those of the real number system. As the preceding section shows, neither the property of continuity nor that of unboundedness is preserved in the number system used by computers. But floating-point arithmetic, the system of numbers used by computers, is also different from the system of real numbers in more subtle ways. Because most computations are now performed by computers and because the system of floating-point numbers that computers use sometimes affects even relatively simple calculations, it is worth examining in more detail how numbers are represented in computers.

At one time, each manufacturer developed its own set of rules for implementing floating-point arithmetic. Today, all manufacturers use essentially the same set of rules. These rules were created by a group of engineers, mathematicians, and scientists working under the auspices of the IEEE. (IEEE was originally the acronym for the Institute of Electrical and Electronics Engineers, an association of electrical engineers. Today, however, the organization's activities are so wide-ranging and its membership so broad that it uses IEEE as its name rather than as an acronym for anything else.) The working group left some—but not very much—flexibility in the way the rules for floating-point arithmetic could be implemented. As a result, the differences among the various floating-point systems in use today are small. What follows is a description of the main parts of the IEEE standard for double precision format, a format that is in common use. ("Double precision" refers to the fact that it uses 64 binary digits or "bits" to represent each number, twice as many as the single precision format and two-thirds as many as triple precision.)

Within the double precision format, all information about any given number must be represented in 64 bits. Because most numbers cannot be represented using 64 bits, a great deal of attention has been given to the efficiency with which the number is stored. The goal is to store as much information about any given number as possible, and the more compact the representation, the less information about the number is lost when it is reduced to 64 bits. While it sometimes helps to think of the more familiar base 10 notation, computers use base 2, and base 2, because it is so simple, allows for certain additional efficiencies.

To begin, imagine that every number x, whether positive or negative, is represented in a binary version of scientific notation:

$x = \pm m \times 2^n$. The 2 shows that the number is being represented in base 2. Both the m, which is still called the mantissa, and the n, which is still called the exponent, are expressed in base 2 as well. As with base 10, the mantissa is still in two parts: The first part of the mantissa, the integer part, is a number that is greater than or equal to 1 and less than the base in which the number is expressed; the second part, the fractional part of the mantissa, is a number that is greater than or equal to 0 and less than 1. In decimal notation, where the base is 10, any of the integers 1, 2, 3, . . . 9 may be used to represent the integer part of the mantissa, but in binary, where all of the digits are either 0 or 1, only the number 1 is needed to represent the integer part of the mantissa. As a consequence, in base 2 there is no need to store the bit that represents the integer part of the mantissa, because it is always 1. In double precision, where 52 bits are used to represent the mantissa, all of them are used to represent the fractional part of the mantissa. The number 1 is added to the mantissa when the number is retrieved from storage and used in computations.

The base is not stored because it is always base 2. The exponent is stored in a surprisingly efficient way. In double precision, 11 bits are allocated for the exponent. The exponent, which is always an integer, varies between 1,022 and 1,023, where these numbers have been expressed in base 10. One can make better use of the 11 "binary storage units" by doing away with the negative sign entirely. This is accomplished by adding –1,023 to each exponent, a strategy that makes every stored exponent positive—every stored exponent is greater than or equal to 1 and less than 2,046. This strategy does away with the need to store the sign. The computer must, of course, subtract 1,023 from the exponent prior to performing any computations or displaying the number for the user. (The two most extreme values, the numbers that bound the range of variation of the exponent, namely 0, which is stored as a string of eleven 0s, and 2,047, which is stored as a string of eleven 1s, are reserved for special cases. These will not concern us here. Exponents less than 0 or greater than 2,047 cannot be stored.)

Now for some arithmetic: Double precision uses 64 bits to represent each number. The mantissa uses 52 bits, and 11 bits are allocated for the exponent. This is a total of 63 bits. One bit is left. The last bit is used to represent the sign of the original number.

As was demonstrated in the preceding section for base 10, most numbers cannot be represented in such a restricted format so that prior to storing any number, it must be rounded to "fit" the slot to which it is assigned. Because we think in base 10, we round in base 10, but this is a little misleading. When a base 10 number is entered into a computer it must be converted to base 2, and this process introduces additional inaccuracies. To see why, consider the number that in base 10 is represented by 0.1. The number one-tenth has a representation in base 10 that terminates after one significant figure, but in base 2, it has a nonterminating representation:

$$1.0 \times 10^{-1} = 1.1001001001001 \ldots \times 2^{-2}$$

On the left side of the equation, a decimal point is used to separate the integer part of the mantissa from its fractional part; on the right, a "binary point" is used to separate the integer part of the mantissa from its fractional part. The fractional part of the binary mantissa consists of the sequence of digits 1 0 0 repeating without ever terminating. Any time that the number 0.1 is entered, therefore, it is rounded and stored as a number that is close to one-tenth but not equal to it, and in the double precision format, only the first 52 binary digits are stored.

Perhaps the most surprising part of this system of floating-point notation is the size of the gaps that occur within the number system. (In what follows, we will revert to base 10 because base 10 is so familiar, but the same reasoning applies to base 2.) As in the preceding section, suppose that 10 slots are used to hold all of the facts about any given number that is expressed in base 10. In the illustrative example on page 58, the first slot is for the sign of the number. The next six slots hold the mantissa. The next slot holds the sign of the exponent, and the remaining two slots hold the exponent.

Now consider how the two numbers 1.23457 and 123,457 are stored in our illustrative system of floating-point arithmetic.

Although they vary considerably in terms of size, they consist of exactly the same digits. Converting to scientific notation, the first number is stored as 1.23457×10^1 or |+|1|2|3|4|5|7 |+|0|1| and the second number is stored as 1.23457×10^5 or |+|1|2|3|4|5|7|+|0|5|, the only difference between these representations occurs in the size of the exponent. Now suppose that the mantissa in each case is increased by the smallest amount possible in this system of floating-point notation. Because only six slots are allocated for the mantissa, the smallest possible increase in the mantissa is 0.00001—that is, the mantissa 1.23457 is increased to 1.23458. Because the mantissa of 1.23457 equals the number 1.23457×10^1, the smallest possible increase in the given number under this method of representing numbers is 0.00001.

Because the number 123,457 also has a mantissa of 1.23457, the same reasoning shows that the smallest possible increase in its mantissa is also 0.00001. But this increase in the mantissa causes the original number to increase from 1.23457×10^5 to 1.23457×10^5, which can also be written as 123,458. Increasing the mantissa of 123,457 by 0.00001, therefore, causes the original number to increase by 1, a difference that is 100,000 times as large as the difference between 1.23457 and 1.23458. This phenomenon is characteristic of floating-point arithmetic: The farther a number is from zero, the larger the gaps between that number and its nearest neighbors.

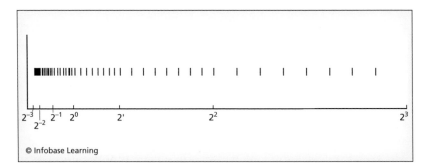

© Infobase Learning

A computer stores only finitely many numbers. The distance between adjacent numbers depends on their size: The larger the numbers the greater the distance between them.

The system of floating-point numbers used by computers has, therefore, little in common with the system of real numbers that students learn in school. Not surprisingly, it is possible to create a calculation where the results obtained by using floating-point arithmetic are very different from those obtained by using the real number system. Despite this, it is often—but not always—the case that the results obtained by computers are similar enough to those that would be obtained using the real number system that the results are still useful. How can this be?

There are three main reasons that the system of floating-point notation used in computers has proven accurate enough to be useful—"useful," that is, from the point of view of numerical computations. First, there is the skill of the programmer. Individuals who do a lot of computations with computers can often recognize when a problem is phrased in such a way that a computer will yield results that are grossly inaccurate, and they take steps to rephrase the problem so that the results that the computer generates will be more accurate. Second, while the computer continually introduces errors, these errors tend to be randomly distributed and so there is a fair amount of cancellation. An error in one direction is often reduced by an error in the other direction. While the user may not know the details of how the various inaccuracies correct for each other, these corrections usually occur, and those who use computers depend upon the existence of such self-correcting cancellations. Finally, in physical problems, there is generally a fair amount of uncertainty about the values of different variables. This uncertainty is summed up in the phrase "significant digits." Not every digit used in a calculation has physical meaning. If there are, for example, five significant digits in the numbers used as input for the computer, and the computer generates an answer that is expressed in 10 nonzero digits, the last five digits of the answer have no physical meaning anyway and can be (and should be) rounded off. The mathematician and physicist John von Neumann, who was also a pioneer in the field of computer science, summed up the situation in these words, "There's no sense in being precise when you don't even know what you're talking about."

Computers have made a huge difference in the quantity and quality of calculations that humans make, and some of these calculations have proven to be very significant. They are used, for example, to improve the safety of aircraft, to smooth the flow of traffic, and to model the workings of the human heart. These calculations are executed so seamlessly that it is easy to forget that they are the result of a great deal of creativity and hard work. For those hoping to make contributions to engineering, science, and some kinds of mathematics, it is important to develop an appreciation for the nature of these calculations if computers are to be used as effectively as possible.

PART TWO

EXTENDING THE IDEA OF NUMBER

5

AN EVOLVING CONCEPT
OF NUMBER

Knowledge of positive fractions and positive whole numbers, what we now call the positive rational numbers, is at least as old as civilization itself. The ancient Egyptians, Mesopotamians, and Chinese all found ways to compute with positive rational numbers. This is not to say that mathematically advanced ancient cultures had number *concepts* similar to ours. Although ancient computational techniques were sometimes identical to those taught in school today, ancient number concepts were often quite different from those with which we are familiar. Nor was the transition from ancient to modern concepts easy or swift. It is an extraordinary fact that the development of a clear understanding of negative numbers took longer than the invention of calculus.

Emotional as well as intellectual considerations proved to be significant barriers to progress. Numbers are fundamental—they are the building blocks and one of the principal sources of examples that mathematicians use as they explore mathematics—and each generation of mathematicians, ancient and modern, is heavily invested in what it assumes numbers represent. Seeing beyond old ideas and preconceptions has always been difficult and time-consuming. This is as true in mathematics as in any other field.

Today we interpret numbers in the language of geometry. At an early age we learn about the "real number line." The idea is simple enough: Imagine a line with two special points. The first point is 0. The 0 point divides the line into positive and negative parts. The second point is the number 1. The distance from 0 to 1 establishes a scale. The point that we associate with the

A page from one of the oldest surviving copies of Euclid's Elements. *The Greek concept of number was very different from ours.*

number 2, for example, is located on the same side of 0 as the number 1 and exactly twice as far from 0 as the 1 point. The position of the number 1 on the line also determines a direction. Numbers on the same side of 0 as the number 1 are positive. Those located on the opposite side of 0 from the number 1 are negative. The position of every other number on the line is now determined. In this representation of the number system, every number represents a *directed distance* from 0; that is just another way of saying that the positive numbers lie to the right of 0, the negative numbers lie to the left of 0, and each number communicates information about how far the corresponding point is located from the origin. This simple and elegant representation of our real number system is a powerful conceptual aid. It is also a relatively recent innovation.

Earlier concepts of number did not depend on the real line. For the most part the numbers that early mathematically advanced cultures used were all positive. Geometrically they used only the numbers that are located to the right of 0 on the real number line. Most early cultures simply refused to consider negative numbers or the number 0. Lacking a real number line, they identified numbers with objects or parts of objects. Sometimes the set of positive whole numbers sufficed: 10 people, 5 lakes, 3 mountains, and so on. Other times fractions were necessary: 3 1/2 feet, 1/4 kilogram, or 1/2 day, for example. For most of human history no one took time to consider a set of no apples, or a collection of –3 days. Fortunately the collection of positive rational numbers was adequate for *most* purposes. We emphasize the word *most*, because there have always been indications that other numbers exist. There were even indications that they might be useful. Mathematically stumbling on other classes of numbers was almost unavoidable. To understand how classes of numbers outside the set of positive rational numbers were discovered, considering the operations that make up basic arithmetic is helpful.

There are four basic operations of arithmetic: addition, subtraction, multiplication, and division. In what follows we need to be concerned with a fifth operation as well, the extraction of roots. (By *extraction of roots* we mean finding square roots, cube roots, and so on.) Three of these operations presented no conceptual problems to early mathematicians. For example, when we add two positive rational numbers the result is another positive rational number. Similarly if we multiply or divide two positive rational numbers the result is still another positive rational number. Today mathematicians describe this situation by saying that the positive rational numbers are closed under addition, multiplication, and division. The word *closed* means that we cannot obtain a number outside the set by adding, multiplying, or dividing numbers chosen from inside the set. The situation is different with subtraction and the extraction of roots.

When we subtract 1 from 2 we obtain a positive rational number. The answer, of course, is the number 1. When we subtract 2 from 1, however, our answer is negative. Subtracting 2 from 1 leads us off the positive half of the real number line. Today we would accept the number –1 as an answer to the problem of subtracting 2 from 1, but earlier in history mathematicians often refused to consider the question. To them it seemed impossible to subtract 2 from 1. Problems that we answer with a negative number, they asserted were unsolvable. There was no mathematical reason for them to restrict their attention to positive numbers. This limitation was one that these mathematicians imposed on themselves. Nor was the only conceptual difficulty the existence of 0 and negative numbers.

In mathematics many more possibilities exist than simply whether or not a number is greater than 0 or less than or equal to 0. If, for example, we search for the square root of 2, written $\sqrt{2}$, we find a number that cannot be expressed as a positive rational number. The number $\sqrt{2}$ *is* positive, but it is not *rational*; that is just another way of saying that it cannot be expressed as a fraction with a whole number in the numerator and another whole number in the denominator. One more example: If we search for the $\sqrt{-2}$, we find a number that has no place anywhere on the real line. The $\sqrt{-2}$ exists, but not in the real number system. Earlier in history mathematicians would have classified the number we now write as $\sqrt{-2}$ as meaningless.

We now know that $\sqrt{-2}$ is not meaningless. It is an example of a complex number. Complex numbers have important uses in science and mathematics. The search for the meaning of different types of numbers has occupied some of the best mathematical minds in history. The search for a number system that is closed under basic arithmetic and the extraction of roots was finally completed less than 200 years ago. Many of its basic properties were discovered even more recently.

Irrational Numbers

An *irrational number* is a number that cannot be written as a quotient of two whole numbers. Roughly 4,000 years ago the Mesopotamians first encountered the irrational number $\sqrt{2}$ when they computed the length of the diagonal of a square with side of length 1. The Mesopotamians were extremely adept at calculation. They had the mathematical acumen to compute highly accurate *rational number* approximations to $\sqrt{2}$: That is, they could find rational numbers that are very close to the irrational number $\sqrt{2}$. Cuneiform tablets have been found with rational approximations to the $\sqrt{2}$ that are accurate to the 1,000,000th decimal place. This is an accurate enough estimate of the size of √2 for most applications—even today.

The algorithm used by the Mesopotamians was *recursive:* This means that it consisted of a series of steps that could be repeated as often as desired. The more often the steps were repeated the more accurate the estimate of √2 became. In particular if we specify that we want an approximation to be within some margin of error— suppose, for example, that we specify that we want an approximation for √2 that is within 1/1,000,000,000th of the true value—the Mesopotamian algorithm can find a rational number that lies within the specified tolerance. All that is necessary is to repeat the series of steps that make up the algorithm enough times.

Theirs was an excellent algorithm for *approximating* √2, but it *cannot* be used to find an exact value for √2. The reason is that each step of the Mesopotamian algorithm consists of a sequence of additions, multiplications, and divisions. The initial input in the Mesopotamian algorithm was always a positive rational number and, as we have already pointed out, the positive rational numbers are closed under additions, multiplications, and divisions. As a consequence *every* sequence of additions, multiplications, and divisions of a set of positive rational numbers can produce only other positive rational numbers. Because the √2 is irrational there is always some distance between the √2 and the number computed with the algorithm.

To understand why the Mesopotamian algorithm can approximate irrational numbers with rational ones, some knowledge of the placement of the rational numbers is useful. Given any two numbers we can always find a rational number between them. It does not matter how close together the numbers are. As long as the two numbers are distinct, there is a rational number that is greater than the smaller one and less than the larger one. Today mathematicians describe this situation by saying that the rational numbers are dense on the real number line: That is, there is no interval, however small, that does not contain a rational number.

The denseness of the rational numbers on the real line helps to explain why the Mesopotamian algorithm for computing √2 works so well. We can imagine surrounding the √2 with a small interval. Geometrically the "smallness" of the interval represents our margin of error. What the Mesopotamian algorithm allows us to do is to calculate one of the rational numbers that lies inside this interval, and there is always a rational number inside the interval because the rational numbers are dense on the real line. The Mesopotamians would not have understood their algorithm in these terms, because they would not have known how the rational numbers are placed on the real number line. Nor is there any indication that they were aware of the existence of irrational numbers. Nevertheless they made use of these facts whenever they employed their algorithm for computing square roots.

Although the Mesopotamian algorithm yielded only an approximation to $\sqrt{2}$, there is no indication that this "shortcoming" was a source of concern to them. Nor is there any indication that they were curious about why each step of their algorithm would yield a better approximation to $\sqrt{2}$ instead of $\sqrt{2}$ itself. Rather than searching for a perfectly exact answer, they simply stopped when their approximation was "accurate enough." The degree of accuracy they required depended on the application that they had in mind.

It may seem strange that they did not investigate the problem of obtaining an exact answer versus an inexact estimate. But if the idea of finding *only* an approximation for $\sqrt{2}$ seems not quite satisfactory, we should remember that our calculators are also capable of finding only rational approximations of irrational square roots. This is a consequence of the way calculators and computers store numbers. In fact the algorithm used by our calculators to compute square roots is often the very same algorithm that was used by the Mesopotamians!

Pythagoras of Samos

As math students we usually encounter positive irrational numbers for the first time when we compute square roots. Historically this was the way they were first uncovered as well. The credit for the discovery of irrational numbers is usually given to the Greek philosopher and mathematician Pythagoras of Samos (ca. 582–ca. 500 B.C.E.) and his followers. The Pythagoreans discovered that the number √2 is irrational. They learned something new about the nature of this number. There is no indication that they discovered anything new about the size of the number. The Mesopotamian estimate of the *size* of √2 was almost certainly more accurate than any estimate that Pythagoras ever knew.

In contrast to the work of the Mesopotamians, the discovery by the Pythagoreans that √2 cannot be represented as a quotient of two whole numbers reveals little about the size of the number. Still it was an important mathematical insight made by one of the most influential of all the Greek philosophers and mathematicians.

Pythagoras was a philosopher and a mystic with a penchant for travel. As a young man he visited Egypt and perhaps Mesopotamia, and he absorbed the philosophy and mathematics of these societies. He eventually settled in the Greek city of Cortona, located in what is now Italy. He was a charismatic person and was soon surrounded by followers. They built a settlement and lived apart from the rest of the community. Pythagoras and his followers lived communally, sharing everything, owning nothing individually. Their philosophy of communalism also applied to claiming credit for new ideas and new mathematical discoveries. As a consequence we cannot distinguish between what Pythagoras discovered as an individual and what his disciples learned on their own. Probably Pythagoras and his disciples would have rejected that kind of he–they distinction anyway. We can, however, be certain of this: It was the Pythagoreans who first learned of the existence of numbers that are not rational: that is numbers that cannot be expressed as the quotient of

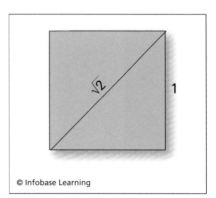

© Infobase Learning

Geometrically the irrationality of √2 means that there does not exist a line segment of length x, *where* x *can represent any number, and whole numbers* m *and* n *such that* mx = 1 *and* nx = √2. *In other words, the side of the square (of length 1) and its diagonal (of length √2) share no common measure.*

two whole numbers. Their discovery disappointed and unsettled them.

Mathematics was a central part of Pythagorean philosophy. "All is number" was their maxim. But when they asserted that all is number they were referring only to a very specific kind of number. They believed that all relationships in nature could be expressed as ratios of whole numbers. When they discovered the existence of irrational numbers, numbers that *cannot* be expressed as ratios of whole numbers, they also discovered that their concept of number and its role in nature was flawed. Because these ideas were central to their philosophy, they tried—unsuccessfully—to conceal their discovery.

The discovery of the irrationality of $\sqrt{2}$ soon led to the discovery of a larger set of irrational numbers. The Greeks soon composed

THE IRRATIONALITY OF $\sqrt{2}$

The proof that $\sqrt{2}$ is irrational is one of the most famous proofs in the history of mathematics. The Greek philosopher Plato (ca. 427 B.C.E.–347 B.C.E.) was so amazed by the existence of irrational numbers that he divided his life into that part that occurred before he saw the proof and that part that occurred after. Plato was not alone. The existence of irrational numbers surprised and fascinated the Greeks.

The proof is a nonexistence proof. To show that $\sqrt{2}$ is irrational we must demonstrate that there are no integers a and b such that $\sqrt{2}$ equals a/b. The proof depends upon two facts about even numbers:

   1. If a^2 is divisible by 2 then $a^2/2$ is even.

   2. If b^2 (or a^2) is divisible by 2 then b (or a) is even.

To show that $\sqrt{2}$ is irrational we begin by assuming the opposite: Suppose that $\sqrt{2}$ can be represented by a fraction a/b where a and b are positive whole numbers. If a and b exist then we can also assume that *the fraction a/b is in lowest terms.* This is critical, but it does not constitute an additional restriction. If, after all, a/b were not in lowest terms, then we could certainly reduce it until it was, so we might as well begin by assuming that a/b is already in lowest terms. Notice that when a/b is in lowest terms, it cannot be the case that both a and b are a short list of numbers that they had proved were irrational, but aside from discovering what irrational numbers are *not*, the Pythagoreans made relatively little headway into discovering what irrational numbers *are*. They knew that irrational numbers *cannot* be represented as a quotient of two integers, but identifying something by what it is not does not yield much insight into the object itself. This type of insight could not have been very satisfying to the Pythagoreans, a group of individuals whose life was entirely devoted to the study of philosophy.

Although the discovery of irrational numbers was important from a philosophical point of view, from a practical point of view the discovery of positive irrational numbers had little effect on computation. The reason, as was previously stated, is that the rational numbers are dense on the real number line. Any computation

---

even numbers. (There is, however, nothing to prevent one of them from being even, but if both are even, then a/b is not in lowest terms.) Here, in modern notation, is how the ancient Greeks reasoned:

Suppose $a/b = \sqrt{2}$
Now solve for b to get
$$a/\sqrt{2} = b$$
Finally, square both sides
$$a^2/2 = b^2$$

This completes the proof. All that is required is to read off what the last equation tells us. The left side is even. (See the first fact.) This shows that b^2 is even. (If one side of the equation is even so is the other.) We conclude that both a and b are even. (See the second fact.) That *both* a and b are even numbers contradicts our assumption that a/b is in lowest terms. We have to conclude that our assumption that $a/b = \sqrt{2}$ is false. This proves that a and b do not exist.

Notice that this proof tells us virtually nothing about the number called $\sqrt{2}$. It is an assertion about what $\sqrt{2}$ is not: It is not rational. Nevertheless this proof is probably one of the most important in the history of mathematics. Irrational numbers continued to perplex mathematicians for the next few thousand years. Their existence was the first indication that numbers could be far more complicated and less intuitive than anyone had imagined.

that involves irrational numbers can be performed with rational numbers, and the difference in the two solutions can be made as small as we please.

Yet the Pythagorean discovery that $\sqrt{2}$ is irrational had important theoretical implications about the existence of a larger, more complete number system. Unfortunately the Pythagoreans did not follow up on their discovery. They continued to work with positive rational numbers. In fact throughout the history of Greek mathematics most mathematicians refused to consider irrational solutions to problems whose solution depended on computation. Diophantus of Alexandria (ca. third century C.E.), the most successful of all Greek mathematicians who were interested in algebra, did not consider irrational solutions acceptable. He posed and solved many problems, but in each case he restricted his attention to solutions that consisted of rational numbers. The discovery of positive irrational numbers hinted at the existence of a larger, more complex number system, but for a very long time no further progress was made in understanding the nature of the number system.

There is one other aspect of the Pythagorean view of number that is different and very important from the point of view of this history of numbers. As previously mentioned, when the Pythagoreans discussed numbers, they meant natural numbers. The Pythagorean emphasis on natural numbers meant that they perceived numbers as discrete quantities. Their number system was, in a sense, comprised only of the natural numbers and the gaps that exist between them. Theirs was a philosophy of the discrete. This is a view of number that is very different from the current one.

Today, because the real number line is introduced at an early age, it is easy to imagine that there are as many numbers as there are points on a line—that is, it is easy to imagine that the set of all numbers forms a continuum, a set without gaps—but this is only one way of perceiving numbers. Other possibilities exist. As a medium for computation, for example, it is not necessary that a number system form a continuum; it is not necessary to take into account the irrational numbers; it is not even necessary that there exist infinitely many natural numbers. As described in chapter 4, for example, computers use only finitely many numbers, none of which are irrational. Because the set of numbers used by any computer is finite, computers use an even smaller set of numbers than that "used" by the Pythagoreans. Since most computations are performed by computers, one could argue that the number system in widest use today is the system of floating-point arithmetic described in the IEEE standard (see page 60), not the system of real numbers.

But the idea of continuity is also important, and it too found a home within Greek mathematics and philosophy just as it has in many cultures that are widely separated by time and space. When the Greeks studied quantities that varied continuously—they called them "continuously varying magnitudes"—they studied geometry. For them, examples of continuously varying magnitudes included lines and planes. The distinction that they maintained between arithmetic, the study of numbers, and geometry, the study of continuously varying magnitudes, was important to them. They viewed these two subjects as opposites.

The Greek perception of numbers as discrete entities outlasted Greek civilization and affected the development of mathematics for many generations and in several cultures. A logically satisfactory description of a continuum of numbers would not be produced until the latter half of the 19th century. The concept of a continuum—described in chapter 8—was motivated, in part, by a desire to fully incorporate the irrational numbers, a Greek discovery, into the number system.

6

NEGATIVE NUMBERS

Whenever a larger number is subtracted from a smaller one, a negative number is the result. Geometrically subtracting a large positive number from a small positive number moves us to the left of 0 on the real number line. Consequently it is hard to avoid occasionally stumbling upon a negative number in the course of solving a problem. This is especially true if the problem requires extensive calculations, because even if the answer is positive, some intermediate steps in the calculation may yield negative numbers. Nevertheless many early cultures did not use negative numbers.

One of the earliest of all cultures to make regular use of negative numbers was the Chinese. In the first section of this volume we discussed the Chinese method of numeration, which used rod numerals. Rod numerals looked as if they had been formed by straight rods, and sometimes that was, in fact, the case. Colored rods, called counting rods, were used as a computation device in ancient China. The user generally carried two sets of rods, one red and one black. The red rods represented positive numbers. The black set represented negative numbers. These were combined according to certain well-defined rules to obtain answers. Such a concrete and elegant system for performing calculations makes the existence of negative numbers easy to accept. Nevertheless the ancient Chinese were cautious. They used negative numbers as intermediate steps during the course of their calculations. They did not accept negative numbers as final solutions to problems. The first great, radical breakthrough in the concept of number occurred in ancient India.

For centuries the astronomical observatory at Ujjain, India, was a center of mathematical research. The observatory consists of many exotic-looking structures. They were used to facilitate precision astronomical observations. (Yanesh Tyagi)

As is the concept of 0, the concept of a number system that fully incorporates negative numbers is an idea we owe to Indian mathematicians. Unlike the concept of 0, which cannot be attributed to any individual mathematician or mathematical text, the first description of negative numbers can be attributed to the writings of the Indian mathematician and astronomer Brahmagupta (ca. 598–ca. 655), who wrote a complete mathematical treatment.

Not much is known of Brahmagupta's life. He was head of the astronomical observatory at Ujjain, which was, for several centuries, one of the great centers of astronomical and mathematical research in the world. His major work is *Brahma Sphuta Siddhānta.* It is principally an astronomy text and was written, in part, for the purpose of correcting and modernizing an earlier astronomy book, *Brahma Siddhānta.* In addition to information about astronomy

Brahmagupta's book contains several chapters on mathematics. In those chapters Brahmagupta explains how to incorporate the number 0 and the negative numbers into arithmetic.

Brahma Sphuta Siddhānta contains little of the stark beauty of Greek texts. Numerous statements are made without proof or other supporting arguments. This may have been the result of the literary form Brahmagupta used to convey his ideas. As was the custom in his time and culture, the book was written entirely in verse. In any case and for whatever reasons, Brahmagupta shows us only his conclusions, which are written in the form of a series of rules. He provides no justification for these rules; nor does he reveal how the rules were discovered.

The concept of number described in *Brahma Sphuta Siddhānta* is a huge step forward. In the section "Algorithm" we find for the first time most of the additional concepts necessary to add, subtract, multiply, and divide positive and negative numbers as we do today. Essentially Brahmagupta expands the set of numbers so that no matter how we add, subtract, multiply, or divide these numbers, we always get an answer that is also in this new and expanded set. Mathematically speaking that is the innovation. Practically speaking his innovation is to expand the set of numbers to fit the problems he wants to solve, rather than to restrict the problems he is willing to consider to those with numerical answers he is willing to accept.

Was Brahmagupta aware of how different his mathematics was from the mathematics that had been developed in Mesopotamia, Egypt, China, and Greece? Perhaps. Indian mathematicians of this era had some awareness of the mathematical accomplishments of the Greeks and possibly of the Mesopotamians and Chinese. Brahmagupta may have appreciated how these new ideas facilitated calculation and how his concept of number made available new solutions to old problems. He does not, however, include any comparisons with other cultures or provide the reader with any explanations that would help us understand how he and his contemporaries had expanded their concept of number.

Here are Brahmagupta's own rules for subtraction. Note that "cipher" is what Brahmagupta calls 0:

The less is to be taken from the greater, positive from positive; negative from negative. When the greater, however, is subtracted from the less, the difference is reversed. Negative, taken from cipher, becomes positive; and affirmative *(taken from cipher)*, becomes negative. Negative, less cipher, is negative; positive *(less cipher)*, is positive; cipher *(less cipher)*, nought. When affirmative is to be subtracted from negative, and negative from affirmative, they must be thrown together.

> *(Brahmagupta.* Algebra with Arithmetic and Mensuration. *Translated by Henry Colebrook. London: John Murray, 1819)*

There are similar rules for the other arithmetic operations. In fact there are several more rules than those to which we are accustomed. (He includes rules for operating with irrational numbers, for example.) His list of rules is followed by a number of problems for the reader to solve.

Although he understands how to perform arithmetic with signed numbers, Brahmagupta encounters problems with the number 0. He clearly recognizes 0 as a number, and he recognizes it as a very special number; notice that in his rules for subtraction, quoted in the excerpt, 0, or "cipher," is always treated separately. But Brahmagupta's insight fails him when he attempts to devise a rule for dividing by 0. Here is his division rule:

> Positive, divided by positive, or negative by negative, is affirmative. Cipher, divided by cipher, is nought. Positive, divided by negative is negative. Negative, divided by affirmative, is negative. Positive, or negative, divided by cipher, is a fraction with that for denominator: or cipher divided by negative or affirmative.

This rule correctly describes how signed numbers should be divided one into another, but he is wrong about 0. Today we know that division by 0 has no meaning. It cannot be defined without introducing logical contradictions. (See the section "Infinity as a Number" in chapter 10 to get some insight into the difficulties involved.) In particular his assertion that 0 divided by 0 equals 0 is false. Moreover he is more than a little vague about the possibility

ANCIENT MATHEMATICAL TEXTS
FROM THE INDIAN SUBCONTINENT

The Hindu mathematical tradition occupies a special place in the history of mathematics. One of the earliest Hindu mathematicians whose work survived to modern times is Aryabhata I (476–550). His book is *Arya Bhateeya* (also called *Aryabhatiya*). Because Aryabhata's work is fairly advanced, it apparently was preceded by many generations of mathematical research. Few records exist before Aryabhata's work, however, and so little is known about the history of mathematics on the Indian subcontinent before he wrote *Arya Bhateeya*.

There is a long history of Indian mathematics after Aryabhata. It is often said that the Hindu mathematical tradition culminated with the work of the mathematician and astronomer Bhaskara II (1114–1185). During the six centuries that separate Aryabhata I from Bhaskara II there are a number of highly creative mathematicians and a number of important mathematical innovations. Despite the breadth and longevity of this tradition of mathematics, however, some characteristics remained the same from generation to generation and distinguish this tradition from other mathematical traditions.

First, in the Indian mathematical tradition mathematics and astronomy are very tightly intertwined. In fact Indian mathematicians produced very few texts on mathematics as a discipline separate from astronomy. Most mathematical writing appears as separate chapters in astronomy texts. Second, in this tradition almost all mathematics is expressed as a formal type of poetry, and that style may account for the terse form of presentation. Third, Hindu texts generally presented mathematical results without proofs. This characteristic is critical. A proof does not just assure the reader that a particular statement is correct. A proof is an aid to the reader who wants to understand the idea behind the statement better. The absence of proofs from most Indian works makes some of the statements found in these texts considerably harder to appreciate. Without a proof the reader who is unsure of what a particular verse means has little recourse other than to read the verse again. Without a proof the mathematician–author who is uncertain of the correctness of a particular statement lacks the ultimate mathematical tool for distinguishing the true from the false. Despite this shortcoming the ideas of the Hindu mathematicians were new and far-reaching, and their results had a profound effect on the development of mathematics around the world.

of dividing a non-0 number by 0. To say that *"Positive, or negative, divided by cipher, is a fraction with that for denominator"* is to say almost nothing. In any case he does not reject the possibility of dividing by 0, which is the modern—and a logically consistent—

OUT OF INDIA

Some but not all Hindu ideas about number eventually found their way out of India. In the ninth century Baghdad was a center for learning and scholarship. There were Islamic, Jewish, and Christian scholars working at the great Islamic institutions of higher education. Mathematicians were, in addition to producing their own research, busy translating the mathematical texts of other cultures into Arabic. Translations of the major mathematical works of the Hindus and Greeks were much sought after. Arab rulers worked to obtain original texts, sometimes from other governments, which were then distributed to Arab scholars for translation and study. It was during this time in this environment that al-Khwārizmī, a mathematician and astronomer living in Baghdad, obtained a copy of a Hindu work. The exact work is no longer known, but some scholars speculate that it was Brahmagupta's *Brahma Sphuta Siddhānta*. Al-Khwārizmī does not indicate his source. In any case he used this text to write an exposition of the Hindu system of numeration. Unfortunately all copies of al-Khwārizmī's book in the original Arabic have now been lost. Only a single Latin translation of al-Khwārizmī's Arabic exposition of a Hindu text, *De Numero Indorum* (Concerning the Hindu art of reckoning), has survived into our own time.

It was al-Khwārizmī's book that introduced the Hindu system of numeration into Islamic culture, and it was through cultural contact with the Arabs that Europeans eventually learned of the Hindu system. Because the Europeans learned of the Hindu system from the Arabs, they attributed the system to the Arabs, although it is clear from the title of al-Khwārizmī's own book that the system is of Indian origin. Al-Khwārizmī was a prominent scholar, whose research into algebra influenced generations of Islamic and European mathematicians, but his conception of number was surprisingly limited given that he seems to have had access to Hindu mathematical texts. Although he recognized the importance of place-value notation, he did not share Brahmagupta's very broad understanding of what a number is. Al-Khwārizmī and other Islamic mathematicians preferred to work with positive numbers only.

solution to the problem. Conceptual difficulties involving arithmetic with 0 were not unique to Brahmagupta, however. Division by 0 continued to pose problems for mathematicians long after the publication of *Brahma Sphuta Siddhānta*. The English mathematician John Wallis (1616–1703), for example, one of the most accomplished mathematicians of his time, asserted that a positive number divided by 0 equals "infinity" and went on to conclude that if we divide that same positive number by –1, it must be "larger than infinite." Brahmagupta's understanding of the arithmetic of signed numbers was not surpassed for about 1,000 years.

The broad Hindu conception of number was, in contrast to the discoveries of the Pythagoreans, extremely useful for computational purposes. To understand the practical value of the Hindu definition of number, comparing the Hindu concept of number with that of the Greeks is helpful. Recall that the Pythagorean discovery of positive irrational numbers had little effect on computation in part because the positive rational numbers are very closely placed on the positive part of the real line. It is always possible to find a positive rational number that is close enough to any given positive irrational number. In other words it is always possible to calculate exclusively with rational numbers and at the same time maintain a negligible error in the results. There is, however, no positive number of any sort "near" the number –100. It is not possible to approximate negative numbers with positive ones reasonably. Brahmagupta's conception of number was not only more abstract; it was also more useful from a computational viewpoint. By incorporating 0 and the negative numbers to form a much larger system of numbers, Brahmagupta and his Hindu contemporaries greatly expanded the concept of solution. They achieved greater freedom in their computations. Later generations of Hindu mathematicians would slightly modify Brahmagupta's work. They would revisit the problem of doing arithmetic with 0, but most of the Hindu conception of number can be found in Brahmagupta's book.

7

ALGEBRAIC NUMBERS

It is difficult to appreciate how slowly a broad conception of number made its way into the European mathematical tradition. For centuries after the dawn of the Renaissance there were many very prominent European mathematicians who simply refused to accept negative numbers as solutions to equations. The French mathematician and philosopher René Descartes (1596–1650) computed negative solutions to certain equations but dismissed them as "false." The Scottish mathematician Colin Maclaurin (1698–1746), a very creative and influential mathematician, stated in his work *Treatise on Fluxions* that negative numbers pose a problem to mathematicians in that they give rise to ideas that lack any physical basis. Despite the widespread aversion to the use of negative numbers, the concept of negative numbers was not—in fact, it could not be—dismissed. Nor was it only negative numbers that these mathematicians had to consider. What we now call complex numbers, numbers that in general have no place on the real line, had been identified early in the 16th century and described as "imaginary." This new class of numbers could not be dismissed either. Both negative and complex numbers arose in the solution of certain common equations. Ideas about equations and the numbers that constituted their solutions evolved together. Each contributed something to the development of the other.

The equations that are most important to the development of the concept of number are called algebraic equations. Algebraic equations are equations of the form $a_n x^n + a_{n-1} x^{n-1} + \ldots + a_2 x^2 + a_1 x + a_0 = 0$, where each x^i is a power of the variable x (so, for example, when $i = 3$, x^3 means $x \times x \times x$) and where each a_i, called the coefficient of x^i, is the rational number by which x^i is multiplied. (The subscript i means that a_i is the number associated with x^i.) The positive whole number n, which represents the highest power of x appearing in the equation, is called the *degree* of the equation. When n is equal to 2, for example, we have an equation of second degree, also called a quadratic equation.

When we refer to a solution of the algebraic equation $a_n x^n + a_{n-1} x^{n-1} + \ldots + a_2 x^2 + a_1 x + a_0 = 0$ we mean a number that, if substituted for x in the equation, yields a true statement. (Remember that the coefficients of the equation are assumed to be known.) Solutions of algebraic equations are commonly called roots. We can also use the term *root* synonymously with the word *solution*.

Second-degree algebraic equations had been known in various forms since antiquity. Mesopotamian, Greek, Chinese, Hindu, and Islamic mathematicians were already familiar with them, and in each of these cultures mathematicians had also studied some equations of degree higher than 2. The Mesopotamians had solved a very narrow class of fourth-degree algebraic equations, but they could not solve most fourth-degree equations. Hindu mathematicians were adept at solving second-degree equations, as were Islamic mathematicians. Moreover Islamic mathematicians developed efficient numerical algorithms for approximating solutions to equations of the third degree. So the challenge of solving algebraic equations was not a new one even though little progress had been made in finding exact solutions during the thousands of years of recorded history before the Renaissance.

Beginning in Renaissance Italy, however, there was a sudden burst of creativity and insight into the exact solution of equations of third and fourth degree. Never before had exact solutions for general equations of the third and fourth degree been discovered. These new algorithms profoundly affected the course of mathematics. This was not because these mathematicians entirely understood what they had discovered. They did not. In fact centuries passed before mathematicians worked out the meaning of these discoveries. Nevertheless the discoveries themselves caused a prolonged burst of mathematical research. The types of numbers that these mathematicians uncovered when they solved equations of the third and fourth degree were surprising. Even more than surprising, the new numbers challenged their conception of what a number is. The new numbers—although many of the mathematicians who computed them refused to accept them as numbers at all—posed a continuing challenge to every mathematician's understanding. Like a pebble in one's shoe, they were impossible to ignore. Their existence demanded an explanation, but for a very long time no satisfactory explanation was forthcoming. The centuries-long struggle to understand how these new types of numbers fit into a broader number scheme resulted in the number system that we know today.

Numbers that are solutions to algebraic equations are called algebraic numbers. New types of algebraic equations introduced new algebraic numbers. For the next few centuries the historical development of our number system proceeded in tandem with mathematicians' understanding of algebraic equations.

Tartaglia, Ferrari, and Cardano

The discovery of new algebraic numbers begins with the Italian mathematician and physicist Niccolò Fontana (1499–1557), better known as Tartaglia. Tartaglia was party to one of the most famous disputes in the history of mathematics. One reason the dispute is so famous is that it was so crass, but another reason is that the source of the dispute, the book *Ars Magna* by Girolamo Cardano, marks for many historians the beginning of modern mathematics. Apart from this dispute, about which we have more to say later, Tartaglia led an active and eventful life. He was born in the city of Brescia, an ancient town located in what is now northern Italy. It was once part of the Roman Empire. In Tartaglia's youth Brescia was one of the wealthiest cities in the region, but the young Tartaglia did not share in that wealth. His father, Michele, a postal courier, died when Tartaglia was still a boy and the family was left impoverished. Not much else about Tartaglia's youth is certain. According to one story Tartaglia, at the age of 14, hired a teacher to help him learn to read. According to the story by the time he

A creative scientist as well as a mathematician, Niccolò Tartaglia is remembered principally for discovering an algorithm that would enable the user to calculate the roots of an arbitrary third degree equation.
(Smithsonian Institution)

reached the letter *k* he ran out of money and from that point on he was self-taught.

In 1512 French forces attacked Brescia and sacked the city. During the battle Tartaglia was severely cut across the face by a sword. The wound left him with a lifelong speech impediment. *Tartaglia* means "stammerer," and for whatever reason, Niccolò Fontana adopted the nickname as his own.

Tartaglia published several books during his life. Some were translations of Greek texts and two were original works. One of his translations was of Euclid's *Elements*. One of his own works, the ballistics treatise *Nova Scientia* (A new science), marks an early attempt to discover the physics of falling bodies. The other book was a three-volume treatise on mathematics, *Trattato di Numeri et Misure* (Treatise on numbers and measures). Today Tartaglia is best remembered for his discovery of a single algorithm. He discovered a method for finding *exact* solutions to algebraic equations of the third degree. His method is, essentially, a formula that uses the coefficients of the equation to compute the roots. Tartaglia's discovery turned out to have many important implications for the history of mathematics, most of which Tartaglia could not have foreseen. He did, however, immediately recognize that he had done something important. At the same time, a professor of mathematics at the University of Bologna, Scipione del Ferro (ca. 1465–1526) had also made some progress in solving third degree equations. He learned to solve a more restricted class of equations of the third degree, but his solution was not widely circulated and he had little impact on the field.

To appreciate what Tartaglia accomplished, recall that mathematicians in several different cultures and times had all used variants of what we call the quadratic formula to find exact solutions for equations of the second degree. No one in history had found an algorithm for generating exact solutions for equations of the third degree. When Tartaglia announced that he had found such an algorithm, his contemporaries were understandably skeptical. A contest was proposed between Tartaglia and a student of del Ferro named Antonio Maria Fior. Tartaglia and Fior each drew up a list of problems for the other to solve. The problems were accompanied by solutions. The solutions were given to a third party, and then Fior and Tartaglia exchanged problems. At the end of the contest Tartaglia had solved all the problems Fior had given him, but Fior had solved none of Tartaglia's. Tartaglia became famous.

Intrigued by news of Tartaglia's discovery, the prominent Italian physician, mathematician, astrologer, and gambler Girolamo

A view of Bologna, Italy—during the Renaissance, Bologna was a center for advanced mathematical research. (EdLab at Teachers College, Columbia University)

Cardano (1501–76) contacted Tartaglia and asked for information about his method. Tartaglia refused. Cardano, however, was not a man who was easily discouraged. (He had, for example, been twice rejected by the College of Physicians in Milan, but he persisted in his attempts to gain admission and was admitted on the third try.) Cardano maintained contact with Tartaglia. He reasoned with him. He badgered him. He tried to bribe him. He was indefatigable. Eventually in a moment of weakness Tartaglia relented and shared the formula with the prominent physician. For his part Cardano agreed never to share the secret with anyone else.

Another important discovery soon followed. Cardano quickly shared Tartaglia's secret with the Italian mathematician Lodovico Ferrari (1522–65). He asked Ferrari to find an algorithm for solving equations of fourth degree, and Ferrari was only too willing to help. Ferrari had been born into a poor family. He initially sought work in Cardano's household as a servant. Cardano, who by this time had become very successful, recognized Ferrari's considerable abilities and took him under his wing. Cardano made sure that Ferrari received an outstanding education. In addition to mathematics Ferrari, with Cardano's help, learned Latin and Greek, the two languages essential to scholars of that era. Ferrari never forgot the debt he owed Cardano. He was fiercely loyal to his teacher and mentor, and when he discovered the algorithm for finding exact solutions to arbitrary algebraic equations of the fourth degree, he did not hesitate to share his discovery with Cardano. In an age when scholars practically worshiped the accomplishments of "the ancients," Tartaglia and Ferrari had done something in mathematics that was truly new.

With both algorithms in hand, Cardano published his best-known book, *Ars Magna* (Great art), which included Tartaglia's famous formula as well as that of Ferrari. That book made Cardano famous as a mathematician. As might be expected, the publication of *Ars Magna* infuriated Tartaglia.

What is important for this history is that the algorithms of Tartaglia and Ferrari eventually resulted in a new concept of number. Some of the numbers that arose as solutions to these equations could not be represented as the lengths of line segments; nor could they be used to identify the number of objects in a set. The complicated algorithmic machinations of Tartaglia and Ferrari indicated the existence of new and not-very-intuitive numbers. Even if one did not accept these numbers as "real"—and in his book Cardano calls them "fictitious"—the numbers could be manipulated according to the rules of arithmetic. More importantly they could be used to generate new and valid results.

Cardano half-recognized that these fictitious numbers have a mathematical existence all their own. Cardano was more than devious; he was also a first-rate mathematician. In *Ars Magna* he shows the reader how to manipulate these new numbers to find new solutions to previously unsolvable problems, although he clearly does not believe that these numbers have any practical utility. For example, at one point he shows that he can find two numbers whose sum is 10 and whose product is 40. It is not difficult to show that if we restrict our attention to numbers that lie on the real line, this problem has no solution. The product of any two *real* numbers that add up to 10 can be no larger than 25. Cardano, however, realizes that this problem has solutions that are not real. He produces the pair of numbers $5 + \sqrt{-15}$ and $5 - \sqrt{-15}$ as solutions to his problem. These numbers are the type of numbers that had become familiar to him through his study of algebraic equations. Notice that there is no real number whose square is -15, so neither $5 + \sqrt{-15}$ nor $5 - \sqrt{-15}$ can be a real number. However, if one multiplies these two numbers together one finds that their product is 40, just as Cardano claimed, and their sum is clearly 10 because the terms involving $\sqrt{-15}$ and $-\sqrt{-15}$ cancel with each other. Cardano is clearly proud of this demonstration, but he cannot get himself to embrace these new numbers. In *Ars Magna* he says, "So progresses arithmetic subtlety the end of which, as is said, is as refined as it is useless."

Part of the difficulty in expanding the set of numbers to include all algebraic numbers is that the extension of the old concept of number required to accomplish this is huge. The jump from the set of positive numbers with which these mathematicians were familiar to the set of algebraic numbers required that they solve several problems simultaneously. We previously mentioned that the set of positive numbers is closed under addition, multiplication, division, and the extraction of roots to the extent that every positive number has a positive root. Initially it may seem that the only problem is the operation of subtraction, because the positive numbers are not closed under subtraction. Simply introducing the negative numbers, however, introduces a new problem. The set of all real numbers is closed under addition, subtraction, multiplication, and division. It is not closed under the extraction of roots: For example, square roots of negative numbers are never real numbers. Introducing the negative numbers solves one set of problems but simultaneously creates another.

For a long time after the discoveries of Tartaglia and Ferrari there was considerable disagreement about which numbers were "real" numbers—real in the sense that they had some foundation in reality—and which numbers were "fictitious" or "imaginary," that is, those numbers that had no basis in reality. Many mathematicians were understandably reluctant to spend their time studying or using numbers that they considered meaningless. Brahmagupta's bold idea of effectively expanding the definition of number to include all numbers that arose in the course of a calculation was not an idea that European mathematicians readily accepted. To use any number generated by a calculation raises—but does not answer—the question, What do these new numbers mean?

Despite these difficulties many mathematicians in the era following Cardano recognized the need to expand their concept of number to include more of the solutions their algorithms were generating. They recognized that without a better, more inclusive definition of number, they could not develop a deep understanding of the algebraic equations that many of them were studying.

Girard and Wallis

The Flemish mathematician Albert Girard (1595–1632) was one of the first mathematicians to adopt a conception of number broad enough to serve as a help rather than a hindrance in the analysis of algebraic equations. Not much is known of Girard as an individual. Different biographies even give slightly different dates for his birth and death. In his own time he was generally described as an engineer rather than as a mathematician, but this is not especially surprising. He had skill in engineering, a talent with which many people can identify, and he was far-sighted in mathematics, a talent with which most people cannot identify. In any case in 1629 in his work *Invention nouvelle en l'algèbre* Girard proposes an extension of the concept of number that far surpasses that of his contemporaries.

First, his understanding of negative numbers is modern. He conceives of negative numbers as directed distances along a line, a clear anticipation of the real number line, and in his solutions to algebraic equations he uses positive, negative, and complex numbers. Nor does he use them only in the course of solving problems. Unlike his contemporaries, he accepts all of these numbers as legitimate roots to algebraic equations. This expanded concept of number allowed him great freedom in the study of algebraic equations.

Second, Girard was interested in the relationship between an algebraic equation and its roots or solutions. Usually we are given an algebraic equation and our goal is to find the roots, but it is possible to work the problem in reverse: Given a collection of numbers can we find an algebraic equation with these numbers as roots? The answer is yes, and Girard made some headway in identifying the relationships that exist between the roots of an algebraic equation and the coefficients that appear in the equation itself. (By contrast if one refuses to accept negative and complex roots then it is impossible to see the relationships that exist between the equation and its roots.)

In the end Girard draws an extraordinary conclusion: He speculates that an algebraic equation has as many roots (solutions) as the degree of the equation. (The *degree* of an algebraic equation is the largest exponent appearing in the equation.) He is right, provided one counts negative numbers and complex numbers as valid roots. This is an important insight into the relationship between algebraic equations and algebraic numbers. Unfortunately he cannot prove that his conclusion is correct. Girard's speculation is not proved until centuries after his death. In retrospect this is not surprising. The name of the result is the fundamental theorem of algebra, and a great deal more mathematics had to be developed before proving the fundamental theorem became possible. Girard's ideas drew little attention at the time he proposed them.

By way of contrast with the ideas of Girard, the French mathematician and philosopher René Descartes's (1595–1650) *La Géométrie* asserted that an algebraic equation can have no more roots than the degree of the equation. This is a much more conservative statement. Descartes's ideas arose out of his unwillingness to expand his concept of number. He was willing to accept fewer roots than the degree of the equation because he could not accept negative and complex roots as valid solutions.

Another prominent mathematician who exhibited insight into numbers was the British mathematician John Wallis (1616–1703). Wallis discovered mathematics fairly late in life. English schools did not emphasize mathematics except as an aid for commerce and the trades. Because Wallis had an upper-class education he did not encounter mathematics before college. Nor did he encounter mathematics in college. He began his adult life as an ordained minister. During this time Britain was engaged in a civil war. Wallis supported the group led by Oliver Cromwell, and when Cromwell's forces captured enemy messages written in code Wallis quickly deciphered them. He earned Cromwell's favor for this accomplishment. A few years later Wallis read his first book on higher mathematics, *The Keys to Mathematics* by William Oughtred. Wallis discovered that he had a talent and a love for the subject, and he tenaciously researched and published his mathematical ideas for the rest of his life.

Wallis proved to be a very creative mathematician. He made contributions to algebra and the branch of mathematics that would soon be called calculus. Isaac Newton claimed that he was inspired to study mathematics after reading Wallis's *Arithmetica Infinitorum*. For the last several decades of his life Wallis had an influence felt throughout the European mathematical community.

In the history of numbers Wallis made an interesting and farsighted contribution. Wallis became interested in geometrically representing the roots of quadratic, or second-degree, algebraic equations. He accepted the idea that algebraic equations have negative and complex solutions. He accepted the idea of representing negative numbers as directed distances along a line. In other words he imagined the positive numbers on one side of 0 and the negative numbers, their mirror image, placed on the other side. The question, then, was where to place numbers that are complex; that is, how could he geometrically interpret numbers that are not positive, negative, or 0?

Wallis imagined the complex numbers that arose as solutions to quadratic equations as the dimensions of squares with negative areas. Since $\sqrt{-1}$ has no place on the real line—there is no real number with the property that if we multiply it by itself the answer is -1—these complex numbers could only be represented in two dimensions. For the number that we would write as $a + b\sqrt{-1}$, Wallis imagined a as a directed distance along what we call the real number line. He imagined b as a directed distance measured in a direction perpendicular to the real number line at the point a. The only role of the $\sqrt{-1}$ is to indicate that b and not a lies off the real line. In this way Wallis associated a picture with each number of the form $a + b\sqrt{-1}$. This was a very insightful way of describing complex numbers, because it associated a plausible interpretation with what most mathematicians of the time considered nonsense. Wallis's conception was as modern as any conception of complex numbers for almost a century, but it did not have much effect on the ideas of his contemporaries.

Wallis's conception of the number system was as forward-looking as any for the next several decades. The German mathematician and philosopher Gottfried Leibniz, one of the preeminent mathematicians of his day, showed a great deal of facility in manipulating complex numbers, but despite his skill, he remained unconvinced about their reality. This was the general attitude toward negative and complex numbers for a very long time. Mathematicians continued to gain expertise in working with the "new" numbers. They found their newly acquired skill useful, because it enabled them to obtain results that they otherwise could not. At the same time most mathematicians worked hard to avoid negative and complex numbers in their answers, because the interpretation that they placed on these numbers caused them to be suspicious of any quantity that they represented. As late as the first half of the 19th century, the prominent and prolific English mathematician Augustus De Morgan wrote, "We have shown the symbol $\sqrt{-a}$ to be void of meaning, or rather self-contradictory and absurd. Nevertheless, by means of such symbols, a part of algebra is established which is of great utility."

Although De Morgan was one of the major mathematicians of his day, he was behind the times in his conception of complex and negative numbers. The tide had begun to turn even before his birth. The Swiss mathematician Leonhard Euler (1707–83) had discovered many important relations among positive, negative, and complex numbers and had learned to incorporate all of these numbers into a theory of functions that produced many important results.

Euler and d'Alembert

Leonhard Euler, the most prolific mathematician in history, originally planned to enter the ministry. His father, Paul, was a minister, and his mother, Margaret, was the daughter of a minister. In addition to being a minister, Paul Euler was something of a mathematician and he helped his son develop his mathematical talent, though apparently both parents hoped that their son would enter the ministry. As is any good son Leonhard Euler was obedient to a point. At the University of Basel he studied theology as well as medicine, languages, mathematics, physics, and astronomy. He took classes from Jacob Bernoulli, one of the great mathematicians of the day, and was a friend of Jacob's sons, Nicolaus II and Daniel Bernoulli. (The Bernoulli family produced many prominent mathematicians, including Jacob, Nicolaus, and Daniel.) Eventually the three young men, Nicolaus, Daniel, and Leonhard, found work at the Saint Petersburg Academy in Saint Petersburg, Russia.

Euler had an extraordinary capacity for mathematics. He wrote on almost every branch of mathematics: groundbreaking research and elementary mathematics books alike. He seemed equally at home with both ends of the mathematical spectrum. It is said that he could perform enormous calculations in his head with little difficulty and that he could sit and write mathematics for prolonged periods without pausing, much as Mozart is said to have written music. Even blindness—and Euler was blind for the last 17 years of his life—had little effect on his output. He continued to dictate mathematics papers right to the end. In the 20th century Euler's collected works, including correspondence, were published by Birkhäuser Publishing of Basel, Switzerland. It is an 82-volume set.

In the field of complex numbers Euler invented much of the symbolism that we use today and he discovered many important relations. He was the first to use the letter i for $\sqrt{-1}$, the notation that is now standard. Euler studied logarithms of negative and complex numbers. He studied expressions such as $(a + bi)^{c+di}$. The number with the decimal expansion 2.71828 . . . was also studied by Euler, and he gave it the name e. This notation has also become standard. (The number e is very important in calculus.) Euler discovered the remarkable equation $e^{i\pi} - 1 = 0$, a result that links the so-called five most important numbers in mathematics in a single equation. All of these discoveries and more require deep insight into what a number is, but Euler's abstract manipulations might also have been dismissed as abstract nonsense had not Euler also begun to discover new and important uses for these numbers.

Euler greatly increased the importance of complex numbers in mathematics. Some of his insights into complex numbers were of a more technical nature: He learned how to extend the definition of many functions that had once been defined only for real numbers so that they were also defined for arbitrary complex numbers. This made these functions more useful. Some of his insights were more down-to-earth. He was able, for example, to use complex numbers in the study of cartography. This is not to say that he was always right when he wrote about complex numbers. He made occasional errors. Nor was he entirely free of the unease that his contemporaries so frequently expressed about the "absurdity" of these numbers. Nevertheless he saw deeply into the nature of complex numbers and greatly expanded their prominence in mathematics.

There was a time in Euler's life when he was friends with the French mathematician Jean Le Rond d'Alembert (1717–83), and it is d'Alembert who made the next important discovery about

THE DEBATE OVER "FICTITIOUS" NUMBERS

What follows are excerpts from the 17th-century text *A Treatise on Algebra* by John Wallis. Keep in mind that Wallis was one of the foremost mathematicians of his time. Notice how he struggles to communicate his ideas about the utility of negative and complex numbers as roots for quadratic equations, the same equations and the same numbers that many students now encounter in high school:

> These imaginary quantities (as they are commonly called) arising from the supposed root of a negative square, (when they happen,) are reputed to imply that the case proposed is impossible.
>
> And so indeed it is, as to the first and strict notion of what is proposed. For it is not possible, that any number (negative or affirmative) multiplied into itself, can produce (for instance) −4. Since that like signs (whether + or −) will produce +; and therefore not −4.
>
> But it is also impossible, that any quantity (though not a supposed square) can be negative. Since that it is not possible that any magnitude can be less than nothing, or any number fewer than none.
>
> Yet is not that supposition (of negative quantities,) either unuseful or absurd; when rightly understood. And though, as to the bare algebraic notation, it import a quantity less than nothing: yet, when it comes to a physical application, it denotes as real a quantity as if the sign were +; but to be interpreted in a contrary sense.
>
> As for instance: supposing a man to have advanced or moved forward, (from A to B,) 5 yards; and then to retreat (from B to C) 2 yards: if it be asked how much he had advanced (upon the whole march) when at C? or how many yards he is

complex numbers. (Euler and d'Alembert later had a falling out.) D'Alembert was a prominent and controversial figure in the world of mathematics. His biological parents were wealthy and socially prominent, but they had no interest in raising d'Alembert. Instead they placed their infant son with a family of modest means by the name of Rousseau. D'Alembert was raised by the Rousseaus, but now forwarder than when he was at A? I find (by subducting 2 from 5,) that he is advanced 3 yards. (Because = 5 − 2 = +3.)

But if, having advanced 5 yards to B, he thence retreat 8 yards to D; and it then asked, how much he is advanced when at D, or how much forwarder than when he was at A: I say −3 yards. (Because 5 − 8 = −3.) That is to say, he is advanced 3 yards less than nothing. . . .

Now what is admitted in lines, must on the same reason, be allowed in plains also.

As for instance: supposing that in one place, we gain from the sea, 30 acres, but lose in another place 20 acres: if it be now asked, how many acres we have gained upon the whole: the answer is, 10 acres or +10. (Because 30 − 20 = 10.) (For the English acre being equal to a plain of 40 perches in length, and 4 in breadth, whose area is 160; 10 acres will be 1600 square perches.) . . .

but if then . . . we lose 20 acres more; and the same question be again asked, how much we have gained in the whole; the answer must be −10 acres. (Because 30 − 20 − 20 = −10.) That is to say, the gain is 10 acres less than nothing. Which is the same as to say, there is a loss of 10 acres. . . .

And hitherto, there is no new difficulty arising, nor any other impossibility than what we met with before, (in supposing a negative quantity, or somewhat less than nothing:)

But now (supposing this negative plain, −1600 perches, to be in the form of a square:) must not this supposed square be supposed to have a side? And if so, what shall this side be?

We cannot say it is 40, nor that it is −40. (Because either of these multiplied into itself, will make +1600; not −1600.)

But thus rather, that it is $\sqrt{-1600}$, (the supposed root of a negative square;)

(Wallis, John. Treatise on Algebra. *London: J. Playford, 1685.)*

he maintained some contact with his biological father. It was from his biological father that he received the money needed to pay for his excellent education. D'Alembert attended Collège des Quatre Nations, where, as Euler did, he studied a little of everything: mathematics, science, medicine, and law. This solid and very broad education was a good investment. Although success occurred early for d'Alembert—he was a member of the Académie des Sciences at the age of 24—he continued to live with the Rousseaus until he was middle-aged.

Today d'Alembert is remembered for several diverse accomplishments. First, he was an insightful mathematician who had a particular interest in the mathematical consequences of Newton's three laws of motion. Second, he is remembered as the editor of an influential encyclopedia. Nor were his duties confined to editing the encyclopedia. He also wrote most of the science articles, a massive undertaking. He is also remembered for an insightful article that he wrote about mathematics and public health with special reference to the disease smallpox.

Of particular interest in this history d'Alembert made an important contribution to our understanding of the number system. Essentially d'Alembert wanted to know whether the set of numbers of the form $a + bi$, where a and b represent real numbers, is "big enough" to include all solutions to all algebraic equations. For example, is the set of numbers of the form $a + bi$ closed under the basic arithmetic operations: addition, subtraction, multiplication, division, and the extraction of roots? More generally he wanted to know whether any combination of these operations when applied to a number of the form $a + bi$ produces another number *of this same form*. For example if we take the cube root of $2 - 3i$, is our answer also of the form $a + bi$? The answer, which is yes, is not at all obvious.

This was an ambitious goal. D'Alembert was, in some ways, attempting to conclude thousands of years of research and speculation about algebraic numbers. The problem had begun more than 2,000 years earlier when the ancient Greeks discovered that the square root of a positive rational number is not necessarily a rational number. For the next two millennia mathematicians had worked to construct a number system big enough to be closed under all the operations of arithmetic and the extraction of roots. D'Alembert wanted to show that if he added, subtracted, multiplied, divided, or extracted a root from a number of the form $a + bi$ the result was another number of the form $a + bi$. He concluded that, in fact, it was. He concluded that the set of all complex numbers is big enough to be closed under these ancient operations.

D'Alembert almost got the proof right. His proof contained some small flaws but these were quickly identified and corrected. Furthermore Euler was able to extend this result to show that many other functions of interest to mathematicians—the logarithmic functions, the trigonometric functions, the exponential function, and so forth—when applied to complex numbers yield numbers that can also be written in the form $a + bi$. In fact the expression $a + bi$ is one way of defining what a complex number is, so another way of describing d'Alembert's and Euler's discoveries is to say that they show that the complex numbers are closed under the operations that they considered. The complex number system turned out to be the minimal extension of the set of positive rational numbers that is closed under any combination of the usual mathematical operations. The problem that had first plagued the Greeks in the year 600 B.C.E. was solved in the 18th century.

It may seem that complex numbers were finally understood, but there was more to be done. Mathematicians were slowly developing a clearer understanding of how real numbers can be geometrically represented as directed distances on a line, but the question remained: How can complex numbers be geometrically represented? The answer really lies in the fact that every complex number is of the form $a + bi$, so that every complex number has a "real" part, which we represent with the letter a, and an imaginary part, the "b part." (Here the

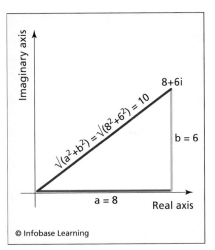

The Argand diagram represents complex numbers geometrically by establishing a one-to-one correspondence between complex numbers and points of a plane. The magnitude, or size, of the complex number 8 + 6i is 10 units, which is the distance from the point (8, 6) to the origin.

letter *i* serves only to distinguish the "real" part of the number from the imaginary part; the imaginary part is the coefficient of the *i* term.) Notice that when *b* is 0, the number *a* + *bi* is a real number. This is an important observation because the real part of the complex number can be represented as a directed distance on the real number line. Likewise the imaginary part can be represented as a directed distance on an "imaginary number line." The answer to the question of how to represent the set of all complex numbers geometrically, then, consisted of setting up a correspondence between the set of numbers *a* + *bi* and the points in the plane, which can be represented by ordered pairs of numbers of the form *(a, b)* (see the diagram on page 103).

Complex Numbers: A Modern View

The first person to have the idea of representing complex numbers geometrically was Caspar Wessel (1745–1818), a Norwegian mathematician, surveyor, and mapmaker. Wessel spent most of his life surveying and making maps, and his discovery, which was published in 1799, appeared as part of a paper on surveying. It attracted no notice from mathematicians until almost 100 years later, when it was rediscovered and republished for the benefit of mathematicians. Wessel's idea, however, had already been rediscovered more than once in the intervening years. In 1806 Jean-Robert Argand (1768–1822), a French bookkeeper, accountant, and mathematician, rediscovered the geometrical interpretation of complex numbers. His name is attached to the usual geometric interpretation of complex numbers, called the Argand diagram. The discovery was made a third time in 1831 by the German mathematician Carl Friedrich Gauss (1777–1855), one of the most prolific mathematicians in history. The Hungarian mathematician János Bolyai (1802–60), one of the founders of non-Euclidean geometry, also discovered something similar. This was clearly an idea whose time had arrived.

Though the idea of representing complex numbers as points on a plane is simple, it is very important to those mathematicians concerned with studying functions of a complex variable.

USING COMPLEX NUMBERS

What we now call complex numbers were initially viewed with a great deal of skepticism. The terminology in use today continues to reflect the old view that complex numbers have no basis in reality. Even now when mathematicians refer to a complex number written in the form *a* + *bi,* they continue to use the old terminology: The letter *a* represents the *real* part of the number, and the letter *b* is the *imaginary* part. The square root of a negative number has no real part and so it is often described as a pure imaginary. It is easy, therefore, to conclude that complex numbers have no applications outside mathematics, but they do. In fact the study of functions of a complex variable is an important part of the education of many engineers and scientists. Here are three common applications:

1. Fluid dynamics: The study of fluid motions—and here the world *fluid* refers to gases as well as liquids—is one of the oldest and most useful branches of all of science. It plays an essential role in the design of aircraft, ships, and certain biomedical devices. The use of complex numbers in the study of two-dimensional flows is about 100 years old, and it is still finding new applications. One of the earliest practical uses for functions of a complex variable in fluid dynamics was in the study of aircraft wing design. The field of complex variables provides a quick and reasonably accurate way for engineers to model the flow of a gas or liquid over a surface. It is a mathematical method that is learned by all undergraduates studying aerospace engineering.

2. Differential equations: The field of differential equations is one of the most important of all branches of mathematics, because the laws of nature are generally expressed as differential equations. In a differential equation the function itself is the unknown. The equation can tell us how the function changes over time, or, how the function changes from one point in space to another. The goal of the mathematician is to discover as much as possible about the unknown function from its differential equation. Methods of solution for these equations often involve functions of one or several complex variables.

(continues)

USING COMPLEX NUMBERS
(continued)

3. Heat conduction: Here the goal is often to compute the temperature of the interior of a body when the only measurements we can make are at the surface. This is important whenever we are trying to predict the reaction of a body to the transfer of heat into or out of it, an important engineering problem. Once again techniques that involve functions of a complex variable play an important role in helping engineers solve these types of equations. These techniques are firmly established—they became important almost as soon as scientists convinced themselves that complex numbers might be useful and they continue to play an important role in this branch of engineering today.

Its importance lies in the fact that a geometric representation of a complex number is a conceptual aid in a subject that even at its simplest is still quite abstract. For example, the distance of any real number to the origin of the real number line is the *absolute value* of the number. The numbers 3 and –3 are both located a distance of three units from the origin. This idea is expressed with the equation $|-3| = |3| = 3$, where the vertical lines represent the absolute value. What, however, is the distance from the origin to the number $2 + 3i$? Once the number $2 + 3i$ is represented as the point (2, +3), the answer is visually obvious. The method of computing the distance is to use the Pythagorean theorem: $|2 + 3i| = \sqrt{2^2 + 3^2}$.

The advantages of using the complex plane, however, extend far past computing absolute values. To appreciate the difficulty of using complex numbers before the invention of the Argand diagram, consider the problem of graphing functions. In the study of functions of a real variable, every student learns to associate a two-dimensional graph with the algebraic description of the function of interest. The graph is important because it establishes a relationship between a geometric and an algebraic interpretation of the function. The picture helps us understand the function better. By contrast the study of functions of a complex variable *begins* at a much higher level of abstraction. The Argand diagram uses two dimensions to represent a single independent complex variable and another two dimensions to represent the dependent complex variable. As a consequence the simplest graphs of functions of a complex variable are four-dimensional. At this level of abstraction any conceptual tool can make a big difference. In fact it was not until mathematicians were able to represent complex numbers as points on a plane that the study of functions of a complex variable really flourished.

8

TRANSCENDENTAL NUMBERS AND THE SEARCH FOR MEANING

The study of algebraic equations helped mathematicians learn about new kinds of numbers. (Recall that an algebraic equation is an equation of the form $a_n x^n + a_{n-1} x^{n-1} + \ldots a_2 x^2 + a_1 x + a_0 = 0$, where each x^i is a power of the variable x—so, for example, when $i = 3$, x^3 means $x \times x \times x$—and where each a_i, called the coefficient of x^i, is the rational number by which x^i is multiplied. The subscript i indicates that a_i is the number that multiplies x^i.) Every root, or solution, of an algebraic equation is called an algebraic number. There are infinitely many algebraic numbers. Every rational number is an algebraic number because every rational number, whether it is positive or negative, is the root of an equation of the form $x + a_0 = 0$. Many irrational numbers are algebraic as well. It may seem that every number is algebraic, but this is not true. It is a surprising fact, about which we have more to say later, that *most* numbers are *not* algebraic. As one might expect, mathematicians were quick to give a name to these nonalgebraic numbers. Perhaps surprisingly the name was chosen before it was shown that any such numbers existed! Numbers that are not algebraic are called transcendental. A *transcendental number* is not the root of any algebraic equation.

Euler suspected that transcendental numbers exist, though he was unable to prove that any particular number is transcendental. Of particular interest to Euler was the number that mathematicians call e. The number e is especially important in calculus and the branches of mathematics that grew out of calculus. Because the number $e = 2.718\ldots$ has an important role in many mathematical and scientific problems, Euler spent a fair amount of effort studying the properties of e. He knew that e is irrational, but he could not show that it is transcendental; that is a much more difficult problem. The difficulty lies in the fact that to show that e or any other number is transcendental, one must show that there is no algebraic equation that has e as a root. In other words one must show that no matter how we choose the degree of the equation, and no matter how we choose the rational coefficients that appear in the equation, when we substitute e for x in the expression $a_n x^n + a_{n-1} x^{n-1} + \ldots a_1 x + a_0$ the result is never 0. This is difficult for any transcendental number. As with Euler, most mathematicians interested in transcendental numbers put their effort into trying to establish the transcendental nature of the two numbers e and π.

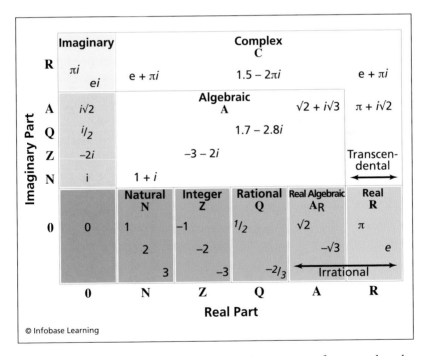

© Infobase Learning

Nineteenth-century researchers established the existence of transcendental numbers: 20th-century researchers showed that almost all numbers are transcendental.

The Swiss mathematician, philosopher, and astronomer Johann Heinrich Lambert (1728–77) was one of the first to gain some insight into the nature of the number π. Lambert must have been an interesting man. He was a member of a family of modest means—his father was a tailor—and from a young age he studied geometry and astronomy on his own. He studied astronomy with instruments that he designed and constructed himself. In fact he was self-taught in all of the subjects in which he excelled. As an adult he earned a living as a bookkeeper and later as a tutor, but his primary interest was always scholarly work. He must have been a very determined scholar. He received recognition for his efforts fairly late in life. In 1768 Lambert published the mathematical result for which he is best known: He showed that π is irrational. This was an important first step.

Toward the end of the 18th century the French mathematician Adrien-Marie Legendre (or Le Gendre; 1752–1833) speculated that π is transcendental, but again neither he nor anyone else at the time could prove that. It was not until about 50 years later that the French mathematician Joseph Liouville (1809–82) proved the existence of transcendental numbers by taking an entirely different tack.

Liouville spent most of his life in Paris teaching and researching mathematics. He showed mathematical promise at a very early age. By the time he was a teenager he was writing papers for mathematical journals. He contributed to many different areas of mathematics. What is important to the history of numbers, however, is that Liouville discovered a set consisting of infinitely many transcendental numbers. All of the numbers that Liouville discovered have the same form. They are all constructed according to the same general formula. We omit his somewhat complicated formula and instead give an example of a number constructed according to Liouville's formula: 0.1100010000000000000000000100 . . . where we place a 1 in the first place to the right of the decimal point, a 1 in the second place (2 = 2 × 1), a 1 in the sixth place (6 = 3 × 2 × 1), a 1 in the 24th place (24 = 4 × 3 × 2 × 1), and so on. Everywhere else we write a 0.

In 1873, almost 30 years after Liouville proved the existence of a set of transcendental numbers, the French mathematician Charles

Hermite (1822–1901), after a great deal of work, completed Euler's project and proved that the number e is transcendental. Nine years after Hermite's discovery the German mathematician Ferdinand von Lindemann (1852–1939) demonstrated that the number π is transcendental.

Part of the difficulty with understanding and using transcendental numbers is that there is no easy way to express them. We can simply write a rational number as a fraction with a whole number in the numerator and another in the denominator. We can also write the algebraic equation associated with an algebraic number that is not rational. This is essentially what we do when we write $\sqrt{2}$. This notation indicates that this number is a root of the equation $x^2 - 2 = 0$. However, we can use neither method with transcendental numbers. A transcendental number is, as a matter of definition, *not* rational and *not* the root of any algebraic equation. Our method of place-value notation is not especially valuable for describing most transcendental numbers, either. If we write out the number as a decimal, we find that the sequence of digits never ends and never forms a repeating pattern in the way that decimal expansions of rational numbers always do. That is why when we refer to the transcendental number formed by the ratio of the circumference of a circle to its diameter—the number that is approximately 3.14—we use the Greek letter π. Similarly the number that occupies a special place in calculus—its value is approximately 2.71—is identified by the letter e rather than by its numerical value. The same shortcoming in our number system explains why we had to describe a procedure for writing Liouville's transcendental number rather than simply writing the number itself. There is no convenient notation for any of these numbers. This is why, although most real numbers turn out to be transcendental, few people, even few mathematicians, can name more than four or five.

It may seem that with the discovery of transcendental numbers mathematicians finally had a firm grasp of the concept of number. They knew something about rational and irrational numbers, positive and negative numbers, real and complex numbers, the number 0, and algebraic and transcendental numbers. They had found

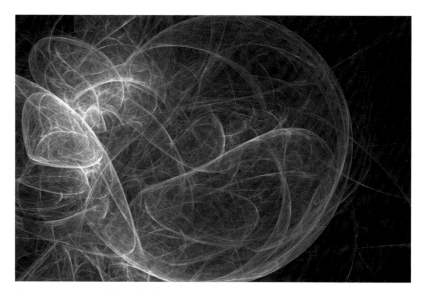

Computer-created form. As mathematicians continue to explore the number systems that they have created, they continue to discover new and often-complex logical structures embedded with these systems.

a number system, the complex number system, that is closed under the operations of arithmetic, the extraction of roots, and other more complicated operations. It may appear that they had a firm grasp of the concept of number, but they did not. Furthermore they knew they did not.

For most of the 19th century mathematicians had worked to understand the nature of various classes of numbers, but there was no unity to their discoveries. As with biologists, who seek to classify the species present in an unfamiliar ecosystem and to identify the relationships between different species, mathematicians had developed a classification scheme for numbers. Each number could be classified as algebraic or transcendental; irrational or rational; positive, negative, or 0; real or complex. But there was no explanation of how, for example, irrational numbers are related to the rational ones. The only definition in use at the time, a definition going back to the Pythagoreans, was that the irrational numbers are not rational, and this was far from satisfying. A deeper insight into numbers was necessary to unify the subject and to ensure a strong logical foundation on which to build the new mathematics of the time. The German mathematician Richard Dedekind (1831–1916) was the first to introduce a unifying concept to the idea of number.

Dedekind and the Real Number Line

Richard Dedekind was a quiet man. He lived with his sister for much of his adult life, and for 50 years he taught high school mathematics. He received a world-class education in higher mathematics at Göttingen University, one of the major centers of mathematics in the world. He entered Göttingen at 19 years of age and received a doctorate three years later. While working as a high school teacher, he made several important and highly original contributions to mathematics. His ideas make up an integral part of the education of every mathematician today.

Dedekind spent most of the second half of the 19th century teaching high school; during that time mathematicians were discovering many aspects of mathematics that were counterintuitive. They discovered properties of infinite sets of numbers that defied their commonsense notions of what should be true, and they were discovering functions that had strange and unexpected mathematical properties. After one such demonstration Georg Cantor, one of the leading mathematicians of the time, said that although he understood the math he still could not believe the outcome. During the second half of the 19th century it became clear that the old ideas and the old techniques were just not powerful enough for the new math. There was a general recognition that the logical underpinnings of the new mathematics were weak. A weak foundation could not be expected to support all of the new mathematics that these mathematicians were attempting to create. Mathematics is a deductive discipline. New results are discovered as logical consequences of previously established ones. If the logic of the existing ideas is weak, the deductions may well turn out to be wrong.

Richard Dedekind's best-known contribution to the foundations of mathematics is his 1872 publication entitled *Continuity and*

Irrational Numbers. In this work Dedekind addressed two conceptual difficulties that had plagued mathematics since the days of Pythagoras. First, he sought to create a positive definition of irrational numbers, a definition that would state what irrational numbers are instead of defining irrational numbers by what they are not (they are not rational). The definition of irrational numbers as numbers that are not rational dates to the Greeks, but as with all such negative definitions it provides little insight into what it purports to describe.

Second, Dedekind wanted to make explicit the continuity of the set of all real numbers. Recall that the Greeks perceived numbers as discrete entities and geometric quantities such as lines and planes as continuous entities. To the Greek mind, the subjects of arithmetic and geometry are opposites in the sense that one deals with discrete quantities and the other with continuously varying ones. By the time of Dedekind, mathematicians assumed that the set of real numbers formed a continuum without being entirely clear about what that assertion meant. By way of example, one prominent scientist of the time, the German physicist and philosopher Ernst Mach (1838–1916), believed that the rational numbers formed a continuum because they were dense on the real number line. (A set is said to be "dense on the real line" if between any two distinct points there is always a third point belonging to the dense set.) But Mach's understanding of the concept of continuum was not shared by Dedekind, who recognized that even though the rational numbers are dense on the real line, there are, nevertheless, gaps between them. Consider, for example, the sequence of rational numbers 1, 1.4, 1.41, 1.414, . . ., where the nth term in the sequence contains the first n digits in the decimal expansion of $\sqrt{2}$. Each term in the sequence is a rational number—the third term, for example, can be written as 141/100—but the sequence converges to $\sqrt{2}$, which is not rational. (The phrase "converges to $\sqrt{2}$" means that if a small circle is drawn about $\sqrt{2}$ on the real line, all but finitely many terms of the sequence will lie within the circle, and this is true no matter how small the circle is drawn provided that $\sqrt{2}$ is contained within it. Sequences in which the terms cluster in this manner are extremely important in mathematics because solutions for many problems are expressed as limits of such convergent sequences.)

Dedekind's demonstration that the set of real numbers forms a continuum rests on the assumption that lines are continuous. That lines are continuous *is* an assumption. Part of what makes Dedekind's article unusual for the time is that he identifies continuity of space as an assumption. He even writes that if space were not continuous, many of the properties that space is assumed to have would remain unchanged. Dedekind's assumption about continuity is not physical or scientific; it is a purely mathematical assumption. What makes Dedekind's ideas valuable are the many interesting and surprising mathematical consequences that follow from it. (See, for example, the section "A New Type of Number" in chapter 12.)

Despite its subtlety, Dedekind's basic idea has a reasonably straightforward description. The idea is to create a correspondence between points on a line and numbers. Imagine a line stretching from left to right. Choose one point and label that point 0 (zero). Choose a second point to the right of 0 and label it with the number 1. Denoting the distance from 0 to 1 as one unit, it is now possible to mark off each point on the line that corresponds to an integer. Those points that correspond to negative integers are located an appropriate number of units to the left of 0, and those points that correspond to positive integers are located to the right of 0. The point corresponding to the integer 7, for example, is seven units to the right of 0. Using techniques known to the ancient Greeks, it is also possible to mark off any point corresponding to a rational number. The distance from one of these "rational points" to 0 is the rational number with which it has been placed in correspondence. Suppose, then, that the correspondence between points on the line and the set of rational numbers has been established.

Now cut the line. (The "cut" is accomplished by choosing a single point on the line. The point determines the location of the cut. The cut may or may not be one of the points that was placed in correspondence with a rational number.) The cut, called *P,* divides the rational numbers into two *disjoint* sets—that is, the two sets share no common elements. If the line is pictured as a string, this is evident. Some of the points lie on the string segment to the left of the cut, and some lie on the string segment to the right of the cut, but no point lies on both. All the rational numbers (or to be more precise all points placed in correspondence with rational numbers) that lie to the left of the cut are less than all of the numbers to the right of the cut, and all of the numbers to the right of the cut are greater than all of the numbers to the left of the cut. Call the set of rational numbers to the left of the cut S_- and the set of rational numbers to the right of the cut S_+. With respect to P, there are three distinct possibilities:

(1) P belongs to S_-, in which case it is the largest rational number in S_-,

(2) P belongs to S_+, in which case it is the smallest element in S_+,

(3) P does not belong to either S_- or S_+.

Notice that if P belongs to S_-, then S_+ does not contain a smallest element. To see why this is so, suppose that S_+ did contain a smallest element. Call it P_1. The point P_1 is greater than P because every element in S_+ is greater than every element in S_- and we have assumed that P belongs to S_-. Choose some other rational number P_2 between P and P_1. It is known P_2 exists because the rational numbers are dense on the line, which is another way of saying that between any two numbers there exists a rational number. (P_2 could, for example, be taken to be the arithmetic average of P and P_1.) Since P_2 is rational and greater than P, it must lie within S_+, but because P_2 is less than P_1, we have the following contradiction: P_1 is *the name given* to the smallest element in S_+. As a matter of definition, no element of S_+ can be smaller. The only way to resolve the contradiction is to conclude that S_+ has no smallest element when S_- has a largest element.

If P belongs to S_+, a similar sort of argument to the one found in the preceding paragraphs shows that when P is the smallest element in S_+, the set S_- has no largest element.

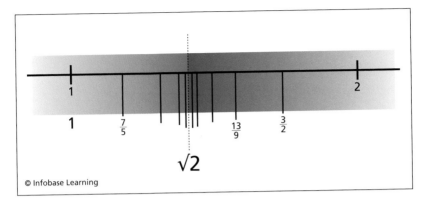

© Infobase Learning

The √2 is the unique point on the real line that divides the set of all points corresponding to rational numbers into two disjoint sets. One set satisfies the inequality a²/b² < 2 and the other set satisfies the inequality a²/b² > 2.

The third alternative is that our cut, *P*, does not belong to either S_- or S_+. Because every rational number belongs to either S_- or S_+, *P* does not, therefore, correspond to any rational number. But a number can be placed in correspondence with *P*. Its value is determined by the way that the cut partitions the set of rational numbers. To determine its value, choose a sequence of rational numbers lying in S_- (or S_+) with the property that the sequence converges to *P* just as was done on page 114, where a sequence of rational numbers was chosen with the property that the sequence converged to √2. The irrational numbers are, then, exactly those numbers that correspond to cuts of the real line that separate the rational numbers into two disjoint subsets with the additional property that the cut does not belong to the set of rational numbers. The value of the number corresponding to such a cut is always determined by the two subsets of the rational numbers that are created by the cut. Today, the cuts are called "Dedekind cuts."

By defining the set of irrational numbers in this way and then appending the set of irrational numbers to the set of rational numbers, Dedekind arrives at a set of numbers with no gaps—that is, every convergent sequence composed of elements from within this set converges to an element belonging to the set. Today this set of numbers, each element of which corresponds to a point on the real line, is called the set of real numbers. Dedekind reasoned that the correspondence that he established between numbers and points on a line demonstrates that the set of real numbers forms a continuum.

Dedekind's construction now provides a positive answer to the question, "What is $\sqrt{2}$?" The answer is that $\sqrt{2}$ is the cut with the property that every rational number a/b on the segment to the left of the cut has the property that $a^2/b^2 < 2$ and every rational number a/b on the segment to the right of the cut has the property that $a^2/b^2 > 2$. The only number with this property is $\sqrt{2}$.

PART THREE

THE PROBLEM OF INFINITY

9

EARLY INSIGHTS

There is nothing in the world around us that prepares us for the mathematical concept of infinity. Everything around us is finite. All of the best science indicates that the universe had a beginning no more than 20 billion years ago, and, one way or another, the universe will end, either quietly or explosively, sometime in the distant future. The universe has a finite life span. Furthermore the universe, as big as it is, contains only a finite amount of mass. Scientists have begun to estimate the amount of mass in the universe. There is still a lot of uncertainty in these measurements, but there is no reason to suspect that they are "infinitely" far off: That is to say, there is no reason to believe that there is an infinite amount of mass in the universe.

Although there is no physical analog for the mathematically infinite, no concept could be more important to mathematics. The idea of infinity pervades mathematical thinking. Infinity makes mathematics possible. Modern mathematics itself did not even begin until mathematicians began to come to terms with the infinite. The first partially successful attempts to deal with infinite sets did not occur until the latter half of the 19th century—surprisingly late. Part of the difficulty in accepting the idea of infinity is that human intuition, honed by a lifetime of familiarity with finite collections (of people, trees, stars, automobiles, and so on), leads us astray when we attempt to formulate even simple statements about an infinite collection of mathematical objects. Applied to the infinite, true statements often sound false and false statements often sound true. Mathematical pioneers who bravely tried to understand the concept of the infinite were often ridiculed

An ultraviolet image of the spiral galaxy M81. As vast as the universe is, nothing in physical space corresponds to our concept of infinity. (NASA/ JPL-Caltech/Harvard-Smithsonian CfA)

for their efforts. Today there is a large body of mathematical work on the properties of infinite sets. The ideas that were generated in the attempt to understand the mathematical concept of infinity have had a profound impact on the development of science and technology. These ideas have helped shape the finite world in which we live.

Philosophically inclined individuals in many different places and times have thought about infinity, but the first culture to attempt to study the infinite in a rigorous way was that of the ancient Greeks. They made only modest progress. The Greek philosopher Zeno of Elea (ca. 490 B.C.E.–ca. 435 B.C.E.) used the concept of infinity in a series of four famous logical arguments. Little is known about

Zeno except that he was a controversial figure. His ideas made quite an impression on the philosophers of his time, and they have resonated among the mathematically and philosophically inclined ever since. The so-called paradoxes of Zeno were made to bolster a certain philosophical position about the nature of the universe. Zeno believed that all of reality is the manifestation of a single being. Because everything is part of a single, eternal, changeless being, change is impossible. Proving the truth of this idea was Zeno's goal, but a positive proof is impossible. Zeno could not prove that he was right, so he did the next best thing: By a clever use of the mathematical concept of infinity he sought to demonstrate that any other worldview leads to a logical contradiction.

Historical sources indicate that Zeno proposed a large number of these paradoxes, but only four have come down to us. These four paradoxes all use some aspect of the infinite to arrive at a conclusion that contradicts our view of reality. One of the arguments, called the Dichotomy, purports to prove the nonexistence of motion. (Motion involves change, and, according to Paramenides, Zenos's teacher, change is impossible.) To appreciate the paradox, imagine a runner just beginning a race. To reach the finish line, the runner must first run half the race and cross the line that is midway between the starting point and the finish line. To reach the midpoint, however, the runner must first run one-fourth of the total distance and cross a point that is halfway between the midpoint of the racecourse and the starting point. This is one-fourth the total distance, but before the runner can run one-fourth of the total distance, the point halfway between a quarter of the distance and the starting point must be crossed. This point is one-eighth the total distance. The pattern goes on and on.

The object of the paradox is to demonstrate that there exist infinitely many "halfway points" between the starting line and the finish line. Furthermore reaching each one of these points must take an "instant" of time. It does not really matter how long an instant is; it is "clear" to most of us that it must take some time to pass each of the points that Zeno has identified. Since it takes an instant to pass each of these points, and since there are infinitely many points, running the race must take an infinitely long time.

(An infinite collection of very short time intervals must still add up to an infinitely long time.) The only possible conclusion, according to Zenos, is that motion is impossible.

The other three paradoxes of Zeno have a similar flavor. Each argument is negative in the sense that it is designed to arrive at a contradiction. Each paradox uses the concept of infinity as a sort of logical sword: Zeno wants to demonstrate that philosophies different from those of Paramenides contain logical fallacies. The only "logical" conclusion, according to Zeno, is to abandon one's own philosophy and adopt that of his teacher. There is no indication that Zeno's paradoxes won many converts for Paramenides. Some philosophers simply dismissed them out of hand as meaningless wordplay. The rest were stymied. Refuting Zeno's paradoxes requires quite a bit of insight into mathematics.

From the point of view of a history of infinity, the Dichotomy is notable in that it fails to convey much information about the nature of infinity. Inherent in the Dichotomy is the assertion that infinitely many instants of time—no matter how small an instant of time is required to pass each point along the race track—add up to an infinitely long time. But what does *infinitely many* mean? How can we distinguish infinite sets from large finite ones? Zeno's paradoxes are of little help.

Perhaps the most important result about infinite sets to come out of the Greek tradition is to be found in the ninth book of Euclid's *Elements*. This work, a multivolume set of mathematics books, is probably the most-read and most-influential work in the history of mathematics. It was written by the mathematician Euclid of Alexandria. We know little of Euclid, except that he lived in Alexandria, Egypt, during the third century B.C.E. We know that his book was written to be used as a textbook and that it contains a careful exposition of mathematics as it was known to the Greek mathematicians of his time. In the ninth book of *Elements* Euclid proves that there are infinitely many prime numbers. The proof that Euclid gives of this fact is remarkable in a number of ways. In fact Euclid's proof that there are infinitely many prime numbers is a wonderful snapshot into the way the ancient Greeks understood the concept of an infinite set. Euclid's writings on this subject offer us an opportunity to see how infinity was perceived by the mathematicians of his age.

To understand Euclid's proof that there are infinitely many prime numbers, one needs to be aware of a few facts about the natural number system. The first fact is a definition: A *prime number* is a natural number that is evenly divisible only by itself and 1. Some examples of prime numbers are 2, 3, 5, and 7. This definition of prime number breaks up the set of all natural numbers greater than 1 into two nonoverlapping classes. Every natural number greater than 1 is either prime or not prime. A number

1	2	3	4	5	6	7	8	9	10
11	12	13	14	15	16	17	18	19	20
21	22	23	24	25	26	27	28	29	30
31	32	33	34	35	36	37	38	39	40
41	42	43	44	45	46	47	48	49	50
51	52	53	54	55	56	57	58	59	60
61	62	63	64	65	66	67	68	69	70
71	72	73	74	75	76	77	78	79	80
81	82	83	84	85	86	87	88	89	90
91	92	93	94	95	96	97	98	99	100

© Infobase Learning

Diagram indicating placement of prime numbers between 1 and 100. Euclid's theorem that there exist infinitely many prime numbers was just the start. Mathematicians continue to study the set of prime numbers today.

that is not prime is called a composite number. (The number 1 is, technically speaking, neither a prime number nor a composite number. It is called a unit.)

The second important fact is about composite numbers. Every *composite number* is, as a matter of definition, divisible by at least one other number besides itself and 1. What the Greeks had already discovered is that every composite number can be divided by some prime number. There are some composite numbers that are also divisible by composite numbers, but *every* composite number must be divisible by at least one prime. From these observations it is easy to see that the prime numbers occupy a special place in our number system: *Every* natural number either is a prime number or is evenly divisible by a prime number.

The last preparatory fact to keep in mind is that no one has ever found a formula that enables the user to list all prime numbers. For many centuries mathematicians have searched for such a formula, a formula that would make sense out of the set of all prime numbers, but the formula, if it exists, remains elusive. It is therefore not possible to use a formula that enables one to "produce" all prime numbers.

By way of contrast we can easily state a formula that allows us to list all even numbers. Imagine that we make a numbered list of all the even numbers. There are many ways of doing this; the most direct way is to list them according to size: Write the number 2 in the first place on the list, the number 4 in the second place on the list, the number 6 in the third place on the list, and so on. In fact we can generalize this scheme and simply say that we write the number $2n$ in the nth place on the list, where we let n represent any natural number—"the number $2n$ in the nth place" is our formula for listing all even numbers. At present if we want to list all primes in the same way that we list all even numbers, we can move down a list of all natural numbers checking each number to see whether there are other numbers besides itself and 1 that evenly divide it. Initially this is easy because the numbers involved are small, but as the numbers become larger and the list of primes becomes longer, the amount of computation involved in determining whether or not a number is prime becomes substantial.

The method we used to show that there are as many even positive integers as there are natural numbers is not suitable for proving that there are infinitely many primes.

This is the situation Euclid faced when he wanted to know whether the set of all primes contains infinitely many numbers. Euclid's approach to this problem is very telling. He does not try to prove directly that the set of all prime numbers is infinitely large. Instead he uses an indirect approach. He begins by *assuming* that the set of all prime numbers contains only finitely many primes. This is his crucial assumption, because *if* there are only finitely many primes and he makes a list of all the prime numbers, then his list, at some point, ends. No matter how he lists the primes—according to size or some other criterion—his assumption that there are only finitely many primes assures that the list must have a last element.

Euclid imagines making such a list. It would have to look something like this: $p_1, p_2, p_3, \ldots, p_n$, where each letter represents one of the prime numbers in the set and p_n represents the last of the primes on the list. (The subscripts refer to the position on the list where each prime appears.) Presumably there are many primes, so the number n, which indicates how many numbers are in the set, is a very large number. The size of n is, however, less important than the fact that it exists at all. Euclid's next step is to "build" a new number. We can call this number M. We obtain M by multiplying all the prime numbers on our list together and then adding 1 to the result: $M = p_1 \times p_2 \times p_3 \times \ldots \times p_n + 1$. This is the proof. All that remains is to figure out what Euclid has done.

Notice that M is bigger than any number on the list of primes, so it cannot be a prime number itself, since by assumption Euclid's list contains *all* prime numbers. The difficulty that arises is that M also cannot be a composite number. To see why, recall that *every* composite number is divisible by a prime number, so if M is composite then it must be divisible by one of those prime numbers. But it is not. If we try to divide M by p_1, for example, our quotient is $p_2 \times p_3 \times p_3 \times \ldots \times p_n$ plus a remainder of 1. This shows that the prime p_1 does not divide M evenly. In fact no matter which prime on our list we choose to divide into M, the result is always the same. We always get a remainder of 1 in our answer. Since every number greater than 1 is either prime or composite—and Euclid has proved that M is neither prime nor composite—he has obtained a contradiction. Because the conclusion is false, the premise must also be false. The premise—that there exist only finitely many primes—must be rejected. Euclid concludes that there are infinitely many primes.

This is a remarkable proof. It is remarkable because it tells us virtually nothing about infinite sets. The proof is constructed in such a way that all one need know about infinite sets is that they are not finite. To be sure, that is not very much insight into infinite sets, but that is all that Euclid needs. He shows only that when he assumes that the collection of all prime numbers is finite he obtains a logical contradiction. Since every set is either finite or infinite, the only possible resolution for the contradiction is that the set of prime numbers is infinite. Euclid's proof is a good illustration of what the Greeks understood about infinite sets: They knew that infinite sets are not finite, but they did not know much else about them.

Euclid's proof also is one of the very first instances of proving a mathematical statement by contradiction. This is a common logical technique today. The mathematician assumes something that is false and then shows that the assumption necessarily leads to a contradiction. If the logic between the assumption and the contradiction is without error then the only explanation is that the assumption made at the outset of the proof is false. Because the assumption is false one can safely conclude that the opposite of the assumption must be true.

Euclid's proof that the set of primes is infinite shows that the Greeks acknowledged the existence of infinite sets. They were, however, never comfortable with the idea of the infinite. They resolved their discomfort by making a distinction between the "potentially infinite" and the "completed infinite." Historically, the distinction is an important one. The Greek preference for the potentially infinite was shared by Islamic mathematicians and most European mathematicians until the 19th century. In some ways, the distinction remains important, and so it is worthwhile to examine the two contrasting ideas of infinite sets.

To understand the idea of the potentially infinite, consider the set of natural numbers. Most people uncritically accept the idea that there are infinitely many natural numbers. But the assertion that there are infinitely many natural numbers implies additional logical consequences that people, once they are made aware of them, are less willing to accept. In particular, the existence of infinitely many natural numbers means that there are natural numbers that are much larger than the sum of all the protons and electrons in the universe—numbers so large that they cannot be generated by any present-day computer or any computer that could possibly be built. In fact, the assertion that there are infinitely many natural numbers implies that "almost all" natural numbers lie outside the range of numbers that humans can use or even imagine and that this will always be true. In a practical sense, these large numbers (assuming that they "exist" at all) are completely unnecessary.

As a practical matter, all anyone requires of the set of natural numbers is that there are enough of them to perform calculations. In other words, from the point of view of calculations, there only need to be enough natural numbers for any specific purpose. This is evidently true for bookkeepers and accountants, but it is also true for mathematicians who use computers to solve very large, computationally intensive problems. They too only require (at most) the concept of the potentially infinite.

The computer programs devised by mathematicians to solve large problems generally use "recursive algorithms," a term that means that calculating a solution involves making several passes. The program will solve part of the problem on the first pass and another part of the problem on the second pass, and so on. Each pass may entail billions or even trillions of calculations. The program is written to make one pass after another until some condition is satisfied at which point the answer is deemed "good enough," and the program is terminated. The idea that the program can cycle as often as is necessary is critical. No matter how many times the program cycles, it can, if necessary, cycle once more. Because each program consists of finitely many operations repeated an unspecified but finite number of times, they make use of the concept of the potentially infinite. They never make use of the completed infinite—that is, they never make essential use of an infinite set.

In the fourth century B.C.E. Aristotle expressed the idea of the potentially infinite and its use in geometry in these words:

> Our account of the infinite does not rob the mathematicians of their science, by disproving the actual existence of the infinite in the direction of increase. . . . In point of fact they do not need the infinite and do not use it. They postulate only that the finite straight line may be produced as far as they wish.

> *(Aristotle.* Physica. *Translated by R. P. Hardie and R. K. Haye.*
> *Oxford: Clarendon Press, 1930)*

By *produced* Aristotle means extended.

The alternative idea, the one explicitly rejected by Aristotle, is called the completed infinite. Operations that involve the completed infinite depend in an essential way on the fact that sets contain infinitely many elements. The completed infinite did not become an essential part of mathematical thought until the latter half of the 19th century because infinite sets have properties that are very different from finite sets and many mathematicians found them difficult to accept. The logical consequences of the completed infinite conflicted with the intuition of most mathematicians when the completed infinite was first introduced into mathematical thought. These 19th-century mathematicians quickly discovered that if they accepted the idea of the completed infinite, they also had to accept the idea that infinite sets come in different sizes—some infinite sets are larger than others—because this is a logical consequence of the completed infinite. They also had to accept the existence of "numbers" that are larger than any natural number—so-called transfinite numbers—and once they accepted transfinite numbers, they had to accept the idea of an arithmetic of the infinite—another logical consequence of the completed infinite. Although these ideas now lie at the foundation of modern mathematical thought, they are of fairly recent origin.

10

GALILEO AND BOLZANO

The first important breakthrough in understanding the nature of infinite sets was made by the Italian scientist, inventor, and mathematician Galileo Galilei (1563–1642). Galileo was the son of the musician and composer Vincenzo Galilei, whose compositions are still occasionally performed and recorded. Growing up, Galileo had very little exposure to mathematics. When he enrolled in the University of Pisa at age 17, he intended to study medicine. While at the university he is said to have overheard a geometry lesson, and this chance experience marks the beginning of his study of mathematics and science.

Before he could graduate from the University of Pisa, Galileo ran out of money and had to withdraw. Whatever he had learned at the university, however, was enough to get him started as a scientist and inventor. Soon he was famous as both, and within four years of dropping out of the University of Pisa he was back—this time as a member of the faculty.

Today Galileo is remembered for several reasons. First, his astronomical observations of the Moon, Venus, and Jupiter changed the way the people of his time perceived their place in the universe. For philosophical and religious reasons most of the people of Galileo's time and place believed that Earth is the center of the universe. Galileo did much to discredit these ideas. Discrediting the teachings of the Catholic Church was, at this time, a very dangerous undertaking. Galileo was threatened with torture because of his ideas. He spent the last several years of his life under house arrest. Galileo is remembered for the grace and courage with which he faced his persecutors as well as for his ideas.

Galileo Galilei. Remembered mostly for his contributions to science, Galileo was also a creative mathematician and the first person to note what is now the defining property of infinite sets: A set is infinite if it can be placed in one-to-one correspondence with a proper subset of itself.

Galileo was also one of the first to combine higher mathematics with experimental physics in a way that is truly modern. His ideas on these matters are contained in the book *Dialogues Concerning Two New Sciences,* which was written near the end of his life. (During this time he was under house arrest, and as part of his punishment he had been ordered to stop writing about science.) In this book Galileo also wrote about the nature of infinite sets.

The observations that Galileo made about infinite sets are fairly brief. He enjoyed mathematics, but he was primarily a user of mathematics—Galileo knew how to employ the mathematics that was known at the time to help him see further into scientific problems. Today we might call him an applied mathematician, because, for the most part, Galileo used mathematics as a tool. Fortunately he also had an eye for interesting mathematical relationships and mathematical ideas. When he spotted an interesting mathematical fact or relationship he did not hesitate to point it out to others, but he did not spend much of his time investigating mathematics as a distinct branch of knowledge. He was, at heart, a scientist.

Galileo's mathematical descriptions in *Dialogues Concerning Two New Sciences* are very wordy by modern standards because his book is a work of fiction. It is a literary work as well as a scientific one, written in the form of a conversation among three characters. The characters, each of whom has a distinct personality, spend four days discussing the ideas on which the new sciences are founded. Each chapter comprises one day in the book. As the title implies, most of the conversation is scientific, but there are nonscientific digressions as well. Some of the dialogue concerns philosophical matters, and some of it is of a mathematical nature.

At one point the characters discuss infinite sets. Through this discussion Galileo describes a very interesting and important property of infinite sets. He seems to have been the first person in history to notice this extraordinary property—at least he was the first person to write about it—a property that plays a very important role in the identification and use of infinite sets. The character Galileo call Salviati who is evidently a stand-in for Galileo himself, explains his ideas about the infinite to the ever-practical Sagredo and their dim-witted friend Simplicio. Salviati makes the discussion as concrete as possible by considering a numerical example. His goal is to prove to his friends that there are as many perfect squares as there are natural numbers. (A *perfect square* is a number whose square root is a natural number. The numbers 4 and 9, for example, are perfect squares, because their square roots are 2 and 3, respectively. The number 5, on the other hand, is not a perfect square, because $\sqrt{5} = 2.236. . . .$)

The assertion that there are as many perfect squares as natural numbers requires proof. It violates commonsense notions. Remember that all perfect squares are themselves natural numbers, but many natural numbers are *not* perfect squares. Mathematically we describe this situation by saying that the set of all perfect squares is a *proper subset* of the set of natural numbers. Galileo's assertion that there are as many perfect squares as natural numbers means that if we remove the set of numbers that are *not* perfect squares from the set of all natural numbers, the set of numbers remaining is the same size as the original set. The only possible conclusion is that for the problem under consideration—and consequently for infinite sets in general—sometimes the whole set is no larger than a part.

From the dialogue of the book, it is clear that Galileo is fascinated by this observation. The main character, Salviati, even goes to some lengths to demonstrate to his friends how few numbers are, in fact, perfect squares. He points out that only a 10th of the numbers between 1 and 100 are perfect squares and only 1 of 1,000 numbers between 1 and 1 million is a perfect square.

Generally the larger the interval he considers, the smaller the percentage of numbers that are perfect squares. So, the simple Simplico wonders, how can it be that there are as many perfect squares as integers?

In the end Galileo's method of proof reduces to making a numbered list. The first position on the list is occupied by the first perfect square, the number 1. The next perfect square on the list—the perfect square in second position—is 4. Next up in third position is the number 9. The general rule is that if we let the letter n represent any natural number, then the perfect square in the nth position on the list is the number n^2. Because Salvati's list of perfect squares never ends—every position on the list is occupied by exactly one perfect square—he concludes that there must be exactly as many perfect squares as there are integers. This is an imaginative and bold conclusion, because it requires us to accept a property that is peculiar to infinite sets, a property that violates our ordinary, everyday experience. *For an infinite set* it is often the case that a part of the set can be placed in one-to-one correspondence with the whole in the following sense: We can pair up a subset with the parent set in such a way that every element in the parent set is paired up with a unique element in the subset and no members of either set are left over. In Galileo's example the "part" consists of the set of perfect squares. The "whole" or parent set consists of the entire set of natural numbers. This is an important observation about infinite sets, and it is well worth our while to explore the idea a little. First, given any infinite set, we can always remove any finite collection of elements without changing the size of the original, parent set. For example, suppose that we remove the first 10 natural numbers from the set of all natural numbers. These 10 deletions leave the size of the set unchanged. To prove this is so, we simply follow Galileo's example and make a numbered list. In first position we write 11, in second position we write 12, in third position we write 13, and in nth position we write $10 + n$. Since every position on the numbered list is occupied by an element from our subset we have proved that the set of all natural numbers greater than 10 is just as large as the set of all natural numbers. Second, it is sometimes even possible, as Galileo himself

demonstrated, to remove infinitely many elements from an
infinite set without changing
the size of the set, although
this must be done with some
care. The property that the
whole can equal the part is
true only for infinite sets. By
contrast for finite sets it is
never true that a proper subset
is the same size as the parent
set from which it is drawn.

Galileo's method for com-
paring the set of natural numbers with the set of perfect
squares also bears mention.
Normally when we want
to compare the size of two
finite sets we count them, but
counting is an ineffective tool
for investigating infinite sets:

1.)	1
2.)	4
3.)	9
4.)	16
5.)	25
.	.
.	.
.	.
n.)	n^2
.	.
.	.
.	.

© Infobase Learning

*Galileo observed that there are as
many perfect squares as natural
numbers, although the set of perfect
squares is a proper subset of the set of
natural numbers.*

Infinite sets are too large. We can try counting the set of natural numbers, but no matter how many we count and no matter
when we stop, most natural numbers remain uncounted. It is to
Galileo's credit that he recognized the importance of establishing
a correspondence between a set whose properties he more or less
understood, the set of natural numbers, and a set, the properties of
which he was less familiar, the set of perfect squares.

The central idea of correspondence is easily illustrated if we
imagine a movie theater just before the lights are dimmed. It is a
simple matter to check for empty seats. If we see empty seats and
there is no one left standing then we know that the set of seats is
larger than the set of people in the theater. If, however, there are
no empty seats and there are a few people standing in the back,
then we can be sure that the number of people exceeds the number
of seats. If all of the seats are occupied and no one is left standing
then there is a one-to-one correspondence between the number

of people and the number of seats. In this case we can be sure that there are exactly as many people as seats. All of our conclusions about the relative numbers of people and seats are valid *even though we may not know either the number of seats in the theater or the number of people*. This is Galileo's insight. It allows him to compare the relative sizes of sets without knowing the true, or "absolute," size of either one. He has no real insight into how large the collection of natural numbers is, but by establishing a correspondence between the set of natural numbers and the set of perfect squares

INFINITY AS A NUMBER

The first serious attempt to work out a mathematics of infinity was made by the mathematicians of the Indian subcontinent. The Indian mathematician and astronomer Bhaskara II (1114–ca. 1185) describes the thinking behind this arithmetic of infinity. (The roman numeral II in Bhaskara's name is used to differentiate him from another prominent mathematician of the same name. We will make no further reference to Bhaskara I.) Bhaskara was one of the preeminent mathematicians in the Hindu tradition. In addition to working as a mathematician he was head of the astronomical observatory at Ujjain, the leading Indian observatory of the time. His writings contain most of what is best in classical Indian mathematics. He demonstrates a thorough knowledge of the algebra that was of interest to the mathematicians of his time as well as an appreciation and mastery of the place-value system of notation, the innovation for which these mathematicians are best remembered today.

Bhaskara is thoroughly familiar with the rules of ordinary arithmetic. He knows how to work with fractions and signed numbers. He says, for example, that the product of two numbers of the same sign is positive and the product of two numbers of opposite sign is negative. He is also somewhat familiar with arithmetic operations involving 0. These ideas had long been part of Indian mathematics, but Bhaskara wants to go further. He attempts to introduce the concept of infinity as a way of "completing" arithmetic. To understand his thinking, let x represent a positive number. Consider the fraction $2/x$. The smaller the chosen value of x is, the larger the fraction $2/x$ becomes. In fact we can make the quotient $2/x$ as large as we want provided we make x small enough. Furthermore for any positive number x, we can always "recover" the numerator by by his listlike approach, he is able to prove that both sets are of equal size.

Galileo's insights are very important to the theory of infinite sets, but Galileo himself does not know what to do next. Having discovered that a proper subset of an infinite set may be exactly the same size as the parent set, Galileo hits a conceptual dead end. He clearly recognizes that he has made an important discovery, but he is unsure about what conclusions can be drawn from it. In the end he concludes that if a proper subset of an infinite set is

multiplying the quotient by x: $x \times 2/x = 2$. Bhaskara wanted to extend these ideas to the case in which x represents the number 0. He begins by asserting that 2/0 is infinity. (In modern notation the idea of infinity is often represented by the symbol ∞; Bhaskara asserts that $2/0 = \infty$.) The reason for this assertion is that as long as x remains greater than 0, $2/x$ represents an ordinary number. The closer x is to 0, however, the larger this quotient becomes. Bhaskara's idea is that as x approaches 0, $2/x$ also approaches a number and he calls this number ∞.

Bhaskara indicates that we can recover the 2 when x equals 0 in exactly the same way that we recover 2 when x is not 0. He asserts that when x equals 0 we need only multiply the quotient by 0 and we get the equation $0 \times 2/0 = 2$, or, to put it another way, $0 \times \infty = 2$. *If this equation were true then we could do arithmetic with the number 0 in just the same way that we do arithmetic with every other number,* but Bhaskara's reasoning is flawed. The problem is that we can repeat every word of this paragraph and simply substitute, for example, the number 3 for 2, and if what Bhaskara says about 2 is true, then what we say about 3 must also be true. In fact any positive number greater than 0 can be substituted for 2, and if the previous statements are true of 2, then they must be true of every other positive number. Let a represent any positive number. According to Bhaskara $a/0 = \infty$, but then, when we try to recover a by multiplying both sides of the equation by 0 we arrive at the result that $0 \times \infty$ could be equal to any positive number. As a consequence $0 \times \infty$ has no meaning. Trying to divide by 0 leads to contradictions, and this is the reason Bhaskara's ideas were rejected. Today mathematicians reject the possibility of dividing by 0. One cannot simply "complete" the number system in the way that Bhaskara envisions. Infinity cannot be treated as if it were an ordinary, finite number.

the same size as the set from which it is drawn, then the ideas of "greater than," "equal," and "less than" have no place in a discussion of the infinite. This is how the character Salviati describes the situation:

> So far as I see, we can only infer that the number of squares is infinite and the number of their roots is infinite; neither is the number of squares less than the totality of all numbers, nor the latter greater than the former; and finally the attributes "equal," "greater" and "less" are not applicable to infinite, but only to finite quantities.

> (*Galileo Galilei.* Dialogues Concerning Two New Sciences. *Translated by Henry Crew and Alfonso de Salvio. New York: Dover Publications, 1954, page 88)*

Galileo's arguments are all correct. Only his conclusion is wrong. We soon see that "greater," "less," and "equal" do, indeed, belong to the study of the infinite.

Galileo's writings on infinite sets had little influence on his contemporaries. For the next few centuries mathematicians continued to shun infinite sets. The next mathematician to make a serious study of the infinite was a little-known Czech priest and mathematician, Bernhard Bolzano (1781–1848). Bolzano graduated from the University of Prague in 1805, when he was ordained a priest and received a Ph.D. in mathematics. He then accepted a position as professor of philosophy and theology at the university. He was a determined and highly creative man. In an age of militarism Bolzano advocated nonviolence and wrote about the futility of war. His ideas were very unpopular with those in authority, both secular and religious, but Bolzano persisted in writing about the necessity of reforming and demilitarizing Czech society despite pressure to desist. In 1819 Bolzano was fired from his position at the University of Prague. He even spent some time under house arrest.

The loss of his position at the university freed Bolzano to devote himself to philosophy and mathematics. His mathematical goals were very ambitious. He wanted to establish the field of mathematics on a more rigorous foundation. In this regard he was very forward-thinking, and today he is generally regarded as a very far-sighted mathematician. During his lifetime, however, his ideas received little recognition.

Bolzano recognized the importance of infinite sets and spent some of his energy studying their properties. Like Galileo he was fascinated by the fact that a set can be put into one-to-one correspondence with proper subsets of itself. Unlike Galileo, who wrote about the natural numbers, Bolzano was interested in what we call the set of real

Bernhard Bolzano. One of the most forward-thinking mathematicians in history, Bolzano's influence on the history of mathematics was modest because he did not publish many of his best ideas. (Dibner Library of the History of Science and Technology, Smithsonian Institution)

numbers. Mathematicians of his time still had only a hazy idea of what a real number is. We can understand Bolzano's ideas about infinite sets of real numbers by using simple graphs.

Consider the interval on the real line consisting of all numbers that are greater than or equal to 0 and less than or equal to 1. This interval is one unit long and contains the points 0 and 1. Next draw a sloping line above this interval. Notice that each point on the line has an x-coordinate no smaller than 0 and no greater than 1. How much the y-coordinates of the points on the line vary depends on the slope of the line. In our picture the y-coordinates begin at 0 and end at 2. This graph shows that we can place all of the points between 0 and 1 into one-to-one correspondence with the set of all points between 0 and 2. This shows that although the segment containing the y-coordinates is twice as long as the segment containing the x-coordinates, there are as many points

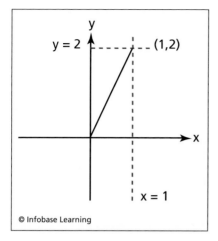

© Infobase Learning

Notice that every x *in the interval [0,1] is paired with exactly one* y *in the interval [0,2], and every* y *in the interval [0,2] is paired with exactly one* x *in the interval [0,1]. This shows that there are as many points in the shorter interval as there are in the longer one.*

on the shorter segment as there are on the longer one. In fact we can make our line slope as steeply as we want. The result is a one-to-one correspondence between the set of points in the interval with endpoints 0 and 1 and the interval with endpoints 0 and n, where we can let n represent any natural number greater than 0.

We can even go further. Imagine a circle. Draw a vertical line through the circle so that it passes through the center of the circle. Label the higher point of intersection of the line and the circle with the letter N and the lower point with the letter S. (See the accompanying diagram.) Now draw a second line. Make the second line horizontal so that it passes through S and is tangent to the circle. (By "tangent" we mean that the line passes through S but no other point of the circle.) Next choose any point on the horizontal line. In the accompanying drawing, this point is labeled P_2. Draw a line containing P_2 and N. Notice that the line intersects the circle below N at the point labeled P_1. This establishes a correspondence between P_1 and P_2.

More generally, every point on the horizontal line, once it is connected to N by a line, determines a unique point on the circle. Similarly, every point on the circle with the exception of N determines a unique point on the horizontal line, which is determined by drawing a line through N and through the second point on the circle until the new line intersects the horizontal line. We have established a one-to-one correspondence between all of the points on the line and a proper subset of the points on the circle, namely the set of all points on the circle except *N*. This demonstrates that there are at least as many points on the circle as there are on the line. (To prove that there are at least as many points on the line as there are on the circle, cut the circle at *N* and "lay it down" along the line.)

We have demonstrated the following: (1) There are at least as many points on the circle as there are on the line, and (2) there are at least as many points on the line as there are on the circle. The only way that both statements can be true is if the two sets of points are equal in size. Because the circumference of the circle is finite and the length of the line is infinite, we have also demonstrated that there are as many points on a finite interval as there are on an infinitely long line. These types of results were considered paradoxical when they were first obtained. But once mathematicians became accustomed to the idea, they were able to create many such demonstrations. The existence of these so-called paradoxes caused many 19th-century mathematicians to become suspicious of the whole subject of set theory.

Bolzano coined the word *set*, and this simple act is also important for the subject. First, with the concept of set, mathematicians had a convenient way of imagining the entire collection of natural numbers. They were all elements in a single set. More important, Bolzano described a *set* as a collection of objects whose definition

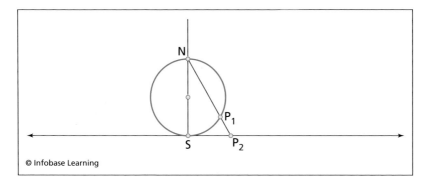

© Infobase Learning

This diagram illustrates the fact that there is a one-to-one correspondence between the points on the real number line and the set of all points on the circle except for the point labeled N. *There are, therefore, as many points on a finite interval as there are on the entire line.*

THE LIFE AND OPINIONS OF
TRISTRAM SHANDY, GENTLEMAN

Sometimes art is better at conveying the concept of the infinite than is mathematics. Although the word *infinite* is familiar to all, the concept is so far beyond our ordinary experience that our intuition is often a poor guide in developing insight into what mathematical infinity means. The mathematician, philosopher, and logician Bertrand Russell illustrated the difference between the finite and the infinite with the tale of Tristram Shandy, and we follow his practice here.

The Life and Opinions of Tristram Shandy, Gentleman, is a novel by the Irish-born English novelist, humorist, and Anglican priest Laurence Sterne (1713–68). *Tristram Shandy* was a radical approach to story-telling for its time. The punctuation, the organization of the story, the stream-of-consciousness style, and the subject matter—the story is the first novel about writing a novel ever written—make the story a remark-able achievement. It was published in nine volumes over a period of eight years (1759–67) and proved to be very popular. The narrator in the novel, Tristram Shandy, has decided to tell the story of his life beginning at the time of his conception. Tristram has a keen eye for detail, and he frequently interrupts his narrative to relate a story about his quirky fam-ily. In fact he has so much to relate to the reader that he does not get

does not depend on the order in which its elements appear. This definition emphasizes how fundamental the concept of correspon-dence is. There are two ways of comparing the sizes of two finite sets: One way is to develop a correspondence between the ele-ments of the two sets; the second way is to count them. Counting a set has significant drawbacks. One disadvantage of counting is that it requires us to introduce an order onto the elements of the set. When we count a finite set we must choose a first element and a last. (In the case of an infinite set there is no last element.) Order is an additional restriction that we impose from without. It has nothing to do with the size of the set. Establishing a corre-spondence allows us to compare sets without introducing an order. Unfortunately none of these important ideas had much influence on Bolzano's contemporaries.

around to describing his birth until volume four of the series. Eventually, however (in volume 9), Shandy realizes that his situation is hopeless: It is taking longer to tell the story of his life than he has time to live it, and the novel ends.

Mathematically speaking the problem Shandy faces is not that he is taking so long to tell the story of his life; it is that he has so little time to tell it. If Shandy had infinitely many years to tell his story, then despite the fact that he continued to fall further and further behind, he would, in the end, have sufficient time to complete the story. It may seem that living longer would only make the situation worse, but this is not the case. Even if it took 1,000 years to relate one year of his life, he would still have enough time to tell the story if he lived forever. After 2,000 years he would have completed the first two years of his life, after 3,000 years he would have competed the third year, and so on. In this manner we can establish a correspondence between each thousand years of writing and the one year of life he purports to describe. The idea is identical to Galileo's correspondence between perfect squares and natural numbers. Our correspondence would look like this: (1, 1,000), (2, 2,000), (3, 3,000),. . . . What is important is that each year of his life would appear somewhere on this list. The fact that he continues to fall further and further behind is irrelevant *if* he has infinitely many years remaining. The longer Tristram lives the more he can tell. If he could live long enough, he could tell us everything.

The investigations of Galileo and Bolzano into infinite sets were similar in concept. We will soon see that the principal difference between the two is that Bolzano investigated sets that are larger than those of Galileo. Both individuals, however, recognized the importance of the idea of correspondence, and both recognized that in the case of infinite sets it is entirely possible that a proper subset can be the same size as the parent set from which it is drawn. Finally, neither Galileo nor Bolzano went much beyond these observations. It would be a few more decades before anyone understood how to found a theory of the infinite based on the observations made by these two pioneers.

11

GEORG CANTOR AND THE LOGIC OF THE INFINITE

The great breakthrough in the theory of sets and in the history of mathematics in general occurred with the work of the German mathematician Georg Cantor (1845–1918). Cantor's insights into the infinite and their importance to the history of mathematics were not immediately appreciated. During much of his life his efforts were ridiculed by many other prominent mathematicians, and he was blocked from a position that he very much wanted at the University of Berlin by one of his former teachers, who was opposed to Cantor's attempts to understand and use infinite sets. Still Georg Cantor did not give up. Today Cantor's work is often described as forming the foundation of the modern age of mathematics.

Cantor was born in Saint Petersburg, Russia, of Danish parents. His family was prosperous, and they remained in Saint Petersburg until 1856, when his father became ill, and they moved to the German city of Frankfurt. During his teen years Cantor showed unusual interest and ability in mathematics. He received most of his education at the University of Berlin, where he was taught by some of the finest mathematicians of the day, including Karl Theodore Weierstrass and Leopold Kronecker. Cantor's doctoral thesis has the engaging name "In Re Mathematica Ars Propendi Pluris Facienda Est Quam Solvendi" (In mathematics the art of asking questions is more valuable than solving problems). Soon after graduation Cantor joined the faculty of the University of Halle, in Wittenberg, Germany, where he remained for his entire working life.

Cantor was not immediately drawn to the theory of infinite sets and their properties. He became interested in the mathematical concept of infinity after discussions with Richard Dedekind, the mathematician who developed a sound logical basis for the real numbers. Perhaps their discussions were entirely mathematical, but interest in mathematics was not his only motivation for his study of the infinite. Cantor's interest in the infinite was also partly theological. Religion must have been an important topic in Cantor's boyhood home. His father was Protestant and his mother was Catholic at a time when this type of "mixed marriage" was both unusual and controversial.

Mr. and Mrs. Cantor, 1880. Although Georg Cantor is now recognized as one of the most important mathematicians in history, his research was often criticized during his lifetime. Many mathematicians found it difficult to accept what Cantor's work revealed. (James T. Smith)

Cantor seems to have believed that insights into the infinite are also insights into God's handiwork. In any case Cantor and Dedekind became lifelong friends, and, when it was possible, Dedekind offered support to Cantor and his controversial ideas.

Cantor's first goal was to identify that aspect that every infinite set shares and that distinguishes all infinite sets from all finite ones. In his proof that the set of all prime numbers is infinite, Euclid simply states that infinite sets are not finite. That is a good enough criterion for the purpose of completing a single proof, but it is not possible to build a comprehensive theory of anything, including infinite sets, based on what it is not.

To begin his theory of infinite sets, Cantor makes a bold decision. He begins his theory by rejecting one of Euclid's most fundamental axioms: "The whole is greater than the part." This simple sentence clearly conforms to our everyday experience. If we spend or consume or tear down part of what we have, whatever remains is less than that with which we began. Before Cantor practically all mathematicians and nonmathematicians alike accepted this idea as a truism. Rejecting Euclid's axiom is necessary, however, because if we use the axiom in the study of the infinite, we cannot prevent logical contradictions. In fact it is precisely the converse of this statement, namely, "There are parts that are equal to the whole," that serves to distinguish all infinite sets from all finite ones. If we accept that for infinite sets there are parts that are equal to the whole, then, instead of encountering logically contradictory consequences, we encounter only strange consequences, and strangeness is no barrier to good mathematics. It is only logically contradictory ideas that must be avoided. Cantor *defines* infinite sets as those sets that have proper subsets that are the same size as the parent set. This definition originates with Dedekind, but Cantor makes greater use of it. The consequences of this definition are disconcerting enough that Cantor's ideas alienated many mathematicians for years.

To state Cantor's definition of an infinite set more formally, we can say that a set is *infinite* provided it can be placed in one-to-one correspondence with a proper subset of itself. Further we call two sets, infinite or not, the same size provided their elements can be placed in one-to-one correspondence with each other. Comparing sets then becomes a matter of finding correspondences between them.

One of Cantor's early discoveries is that the set of all rational numbers is the same size—or, as mathematicians say, "the same cardinality"—as the set of natural numbers. (Two sets have the same *cardinality* if they can be placed in one-to-one correspondence with each other.) That the rational numbers are no more numerous than the natural numbers must have been something of a surprise. If we imagine the natural numbers and the rational numbers as points on the real line, it is easy to see that between any two natural numbers there are infinitely many rational numbers. In fact between any two distinct points on the real line *no matter how closely together they are spaced* there are infinitely many

THERE ARE NO MORE RATIONAL NUMBERS
THAN NATURAL NUMBERS

The fact that there are only as many rational numbers as there are natural numbers surprises many people even today, but the proof of this statement is neither long nor hard. To demonstrate how this is done, we show that there are no more positive rational numbers that there are natural numbers. (The proof is easily extended to show that there are no more positive and negative rational numbers than there are natural numbers.) To establish the correspondence we let a/b represent some rational number, where a and b are natural numbers.

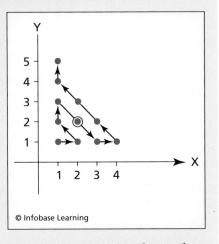

© Infobase Learning

Diagram accompanying the proof that there exist only as many rational numbers as natural numbers

Of course there is more than one way to write a rational number. For example, 1/2 and 2/4 represent the same number, so we always assume that our rational number is in lowest terms. To each positive rational number there corresponds a point on the plane of the form *(a, b)*, where *a* and *b* are positive whole numbers, and a/b is in lowest terms. Similarly, to each point *(a, b)*, where *a* and *b* are positive whole numbers, there corresponds a unique rational number, c/d, such that a/b = c/d and c/d is in lowest terms. (In what follows, we have to be careful to avoid fractions that are not in lowest terms.)

To obtain the necessary correspondence, imagine taking a path through the upper right quadrant of the plane. Our path takes us through each point with positive whole number coordinates. These are called the lattice points of the plane. We use some of these lattice points to form a numbered list. If a point represents a fraction in lowest terms—the point (1, 2), for example, represents the fraction 1/2—then we add it to our list. If the point does not represent a fraction in lowest terms—for example, the point (3, 6) does not represent a fraction in lowest terms—then we do not add it to our list. Those points that are

(continues)

THERE ARE NO MORE RATIONAL NUMBERS
THAN NATURAL NUMBERS
(continued)

added are added in the order in which we encounter them as we travel along our path.

Here is the correspondence: Beginning at the point (1, 1) in the accompanying illustration, follow the path indicated by the arrows. The first point we encounter is (1, 1), and we place 1/1 in position one on our list. The second point is (2, 1); we write 2/1 in position two on our list. Similarly, the third lattice point is (1, 2). We write 1/2 in position three on the list. The fourth point, (1, 3), is likewise added to the list as 1/3 in position four; but the fifth point, (2, 2), is not added to the list because it represents the same rational number represented by the point (1, 1). The difference is that the point (2, 2) is not in lowest terms. We continue to list those lattice points that represent rational numbers that are in lowest terms, and we continue to ignore those lattice points that do not represent rational numbers in lowest terms. By following this path we establish a one-to-one correspondence between the natural numbers, or entries on our list, and the set of all positive rational numbers. Every positive rational number is represented by a lattice point, and given any lattice point, our path eventually passes through that point. If the point represents the rational number in lowest terms, we add it to the list; otherwise, we ignore it. This correspondence is the proof that there are only as many positive rational numbers as natural numbers. This famous result is eventually encountered by all serious students of mathematics.

rational numbers. It appears, therefore, that there ought to be more rational numbers than natural numbers, but there are not. The proof is not difficult (see the sidebar), but other, more subtle results follow. Cantor soon discovers that the set of real algebraic numbers—that is, the set of all real numbers that are roots of polynomial equations with rational number coefficients—is also the same size as the set of natural numbers. This is even more surprising, because the algebraic numbers encompass all the rational numbers and many irrational numbers as well, for example, $\sqrt{2}$ and $^3\sqrt{5} + \sqrt{7}$. In fact the set of all algebraic numbers encompasses almost all the numbers with which we are familiar. The two major exceptions are π and e, which are transcendental. (The letter e represents the irrational number 2.71828 . . . , a number that has a very important role in calculus.)

Given the discovery that the set of rational numbers and the set of natural numbers are the same size, it may seem that Galileo is correct in his assertion that "equal," "greater," and "less" are not applicable to infinite, or, at the very least, that "greater" and "less" are not applicable to infinite sets. But Galileo is wrong. Cantor shows that the set of real numbers is a larger set than the set of all natural numbers. Like Euclid's proof that the set of all prime numbers is infinite, Cantor's proof that the set of all real numbers is larger than the set of all natural numbers begins by assuming the converse, namely, that there exists some one-to-one correspondence between the set of all natural numbers and the set of all real numbers. Cantor's proof shows that no matter what correspondence he chooses, at least one real number does not make it onto the list. The contradiction shows that our assumption that a one-to-one correspondence exists is in error. The set of real numbers is larger than the set of natural numbers. See the sidebar on page 150 for details.

The proof that the set of real numbers is larger than the set of natural numbers has some surprising consequences. To appreciate one such consequence, recall that one can show that the set of all real algebraic numbers can be put into one-to-one correspondence with the set of all natural numbers. (Recall that a number is algebraic if it is the root of an algebraic equation.) The set of numbers that are not algebraic is called transcendental, and so most numbers must be transcendental numbers; otherwise, the set of real numbers is no larger than the set of natural numbers. So although mathematicians can explicitly list relatively few transcendental numbers, it must be the case that "practically all" numbers are transcendental.

Although Cantor demonstrates that infinite sets of different sizes exist, the proof itself raises several interesting and important questions: Are there sets larger than the set of real numbers? How many "degrees of infinity" are there? Can these insights into infinite sets be used to create an arithmetic of infinite—technically called transfinite—numbers? The key to answering some of these questions involves the use of power sets.

THERE ARE MORE REAL NUMBERS THAN NATURAL NUMBERS

The following proof that the set of real numbers between 0 and 1 cannot be placed in one-to-one correspondence with the set of natural numbers is similar to the one originally proposed by Cantor. Because there are at least as many real numbers as there are real numbers between 0 and 1, this proof also shows that the set of real numbers is larger than the set of natural numbers.

The first step in the proof is to express every real number between 0 and 1 in its base 10 decimal expansion. The number 1/3, for example, is written as 0.333 . . . and the number $\sqrt{1/2}$ is written as 0.7071. . . . In fact, any number that has no non-0 digits to the left of the decimal point and at least one non-0 digit to the right of the decimal point represents some real number between 0 and 1. (There is one exception: the number 0.999 . . . consisting of all 9s. This is just another way of writing the number 1, but this makes no difference to our proof.) Next we assume that the set of real numbers between 0 and 1 can be placed in one-to-one correspondence with the set of natural numbers. If the assumption is true, it is possible to make a numbered list, and next to each natural number there is a real number. Moreover every real number between 0 and 1 appears somewhere on our list. The list might look something like this:

1. 0.333 . . .

2. 0.7071 . . .

A *power set* of a set S is the set of all subsets of S. Consider, for example, the set consisting of two letters, *{a, b}*. The power set of *{a, b}* has four elements: *{{a}, {b}, {a, b}, ø}*, where the symbol ø represents the empty set, the set with no members. (The empty set is always a part of every set.) The power set of any nonempty set is larger that the set itself. This is more or less obvious for a finite set, since the power set must contain all single-element subsets of the original set as well as the empty set, but it is also true of infinite sets. If S represents an infinite set, then the power set of S always has a greater cardinality than the parent set from which it is derived. In other words we can

3. 0.10101 . . .

4. 0.66666 . . .

and so forth. (Of course, there is no need to begin with 0.333. . . . We do so only to provide a concrete example. The following proof shows that no matter how we order the set of real numbers between 0 and 1 at least one such number fails to make it onto the list.)

To complete the proof we simply construct a real number that is larger than 0 and less than 1 and is not on this list. We construct our number by using only the digits 5 and 6. We begin our number with a 6, namely, 0.6. This ensures that our number is not equal to the first number on the list, which is 0.333. . . . The next digit is also a 6. This ensures that our number is different from the second number on the list in the second decimal place. At this point our number is 0.66. At the third step we add another 6; this time in the 1,000ths place so that we have 0.666. This third step ensures that our number differs from the third number on the list in the third place. In the fourth place we write a 5. This guarantees that our constructed number differs from the number in the fourth place on our list in the fourth decimal position. The process continues indefinitely. In the nth decimal place we write a 6 provided that the nth decimal digit of the nth number is not 6. If the nth decimal digit is 6, then we write a 5.

The number constructed in this manner does not appear anywhere on the list; it is, however, a real number between 0 and 1. This shows that our assumption that we have a correspondence between the set of all real numbers between 0 and 1 and the set of all natural numbers is in error: No such correspondence exists.

never establish a one-to-one correspondence between any set and its power set.

Cantor shows that the power set of the natural numbers is the same size as the set of real numbers. Furthermore because the power set of the real numbers cannot be placed in one-to-one correspondence with the set of all real numbers—no set has the same cardinality as its power set—the power set of the real numbers is an even larger set than the set of real numbers. The power set of the power set of real numbers is still larger. The concept of the power set enables Cantor to show that there exist not one or two but infinitely many different sized infinite sets. He assigns a symbol to represent the size of each set. These symbols are examples of what are called cardinal numbers.

The cardinal numbers associated with infinite sets are a generalization of ordinary (finite) numbers, and understood correctly they are a perfectly ordinary generalization. To understand the idea of transfinite numbers, thinking seriously about what an ordinary number is can be helpful. There is a difference, for example, between an instance of the number 3 and the number 3 itself. We are surrounded by instances of the number 3: A set of three lakes, a set of three cars, and a set of three women named Sue are all instances of the number 3. These sets are the same in the sense that we can choose any two of them and place the elements of these sets in one-to-one correspondence with each other. The correspondence is, in fact, the only characteristic that all instances of the number 3 have in common. Another way of describing this characteristic that all instances of a given number share is to apply the term *cardinality* to finite sets as well. Let S and T represent any two sets, finite or infinite. The cardinality of S, written $Card(S)$, tells us how many elements are in the set. The sets S and T are both instances of the same number if $Card(S) = Card(T)$; otherwise they are instances of different numbers.

The chief advantage of this approach is that it can be used word for word in describing "transfinite" numbers as well as finite ones. *Transfinite numbers* are numbers that represent the size of infinite sets. The natural numbers, the set of perfect squares, the set of rational numbers, and the set of algebraic numbers are all instances of the transfinite number that Cantor calls $\aleph_0$ (aleph null). So, for example, we can say that $Card([1, 2, 3, \ldots]) = \aleph_0$. The second transfinite number, represented, for example, by the set of all real numbers and by the power set of the rational numbers, he calls $\aleph_1$ (aleph 1). By simply taking the power set of a set that is an instance of $\aleph_1$, he is able to obtain an instance of $\aleph_2$, and so on.

This is a new type of extension of the number system. Cantor has moved from finite numbers to transfinite numbers. He begins to inquire about the possibility of an arithmetic of transfinite numbers. He is successful in generalizing addition and multiplication into something that is now called transfinite arithmetic.

"The arithmetic of the infinite" may, at first, seem difficult to imagine. As in so much of mathematics, however, the key to new ideas and new concepts lies in a reexamination of the familiar. To appreciate transfinite addition, we need to see how it is similar to ordinary addition. Let the plus symbol + represent the operation of union of sets as well as its usual meaning of addition. Let A and B represent two sets. Then $A + B$ represents the set that is composed of all the elements of A and all the elements of B. For example if we let A represent the set $\{a, b\}$, and B represent the set $\{p, q, r\}$ then the set $A + B$ represents the set $\{a, b, p, q, r\}$. (In what follows we will always assume our sets share no elements in common.)

Now consider the very simple equation $2 + 3 = 5$. We learned this equation early in grade school; now we only have to restate the equation in the language of sets. Let the set A be an instance of the number 2, for example, $A = \{a, b\}$, and let the set B be an instance of the number 3, for example, $B = \{p, q, r\}$. To understand Cantor's transfinite arithmetic, we just reinterpret $2 + 3 = 5$ as $Card(A) + Card(B) = Card(A + B)$. Now let x and y represent any numbers, finite or transfinite. Let X and Y represent instances of the cardinal numbers x and y, respectively. First we have $x + y = Card(X) + Card(Y)$. Because $Card(X) + Card(Y) = Card(X + Y)$, we conclude that $x + y = Card(X + Y)$: To find the sum of x and y we need only determine the cardinality of the union of X and Y. In this way Cantor extends the idea of addition from finite numbers to transfinite ones.

Multiplication can also be generalized to transfinite numbers. The inspiration for doing so also arises from reexamining the ordinary process of multiplication of ordinary numbers. To do this we use the multiplication symbol, ×, in a new way. Consider the following example: Let A and B again denote the sets $\{a, b\}$ and $\{p, q, r\}$, respectively. Define $A \times B$ to be the set of all ordered pairs that can be formed by using an element of A for the first coordinate and an element of B for the second coordinate. In this case $A \times B$ represents the following set: $\{(a, p), (a, q), (a, r), (b, p), (b, q), (b, r)\}$. Notice that when A has two elements and B has three elements $A \times B$ has six ($= 2 \times 3$) elements. It is just as easy to show that $B \times A$ also has six elements. This gives rise to the equation $Card(A) \times$

$Card(B) = Card(A \times B)$, and this equation is taken to hold for every pair of sets, finite or infinite. Let x and y be two numbers, either transfinite or finite. Let X and Y be instances of the numbers x and y. To multiply x by y we use the equation $x \times y = Card(X) \times Card(Y)$, and since $Card(X) \times Card(Y) = Card(X \times Y)$ we have our answer: $x \times y = Card(X \times Y)$.

Subtraction and division cannot be defined. The expression $\aleph_0 - \aleph_0$, for example, cannot be given any meaning. Neither, for example, can $\aleph_0 \div \aleph_0$. Nevertheless Cantor is able to imagine an arithmetic of the infinite that is, where it applies, an extension of ordinary finite arithmetic.

As with all good ideas, Cantor's ideas raise many questions, some of which have not been answered. In particular Cantor's discovery of transfinite numbers leaves open the possibility that there are other, not-yet-discovered transfinite numbers. Cantor has demonstrated the existence of a sequence of transfinite numbers, that is, $\aleph_0, \aleph_1, \aleph_2 \ldots$, but are there other transfinite numbers in addition to these? These "other" numbers have to have the property that they fit in between Cantor's transfinite numbers. For example, does there exist a transfinite number that is strictly bigger than $\aleph_0$ and strictly smaller than $\aleph_1$? In this case an instance of this in-between number is too big to be put into one-to-one correspondence with the set of natural numbers, and too small to be put into one-to-one correspondence with the set of real numbers. This problem, called the continuum hypothesis, is one of the

+	n	$\aleph_0$	$\aleph_1$
n	2n	$\aleph_0$	$\aleph_1$
$\aleph_0$	$\aleph_0$	$\aleph_0$	$\aleph_1$
$\aleph_1$	$\aleph_1$	$\aleph_1$	$\aleph_1$

×	n	$\aleph_0$	$\aleph_1$
n	n^2	$\aleph_0$	$\aleph_1$
$\aleph_0$	$\aleph_0$	$\aleph_0$	$\aleph_1$
$\aleph_1$	$\aleph_1$	$\aleph_1$	$\aleph_1$

© Infobase Learning

An addition table and a multiplication table that include two transfinite numbers. The letter n *represents an arbitrarily chosen natural number, and* $\aleph_0$ *and* $\aleph_1$ *represent the first two transfinite numbers.*

famous problems in mathematics. It is easily generalized by asking whether there are other transfinite numbers that lie between each transfinite number in the sequence $\aleph_0, \aleph_1 \aleph_2, \ldots$. This problem is called the generalized continuum hypothesis. These problems lie at the very foundation of mathematics.

Cantor's study of the infinite led to deep insights into the nature of infinite sets. It also led to new interpretations of the most elementary of arithmetic operations. Although his ideas remained controversial for some time, Cantor's set theory was soon embraced by some of the foremost mathematicians of his age because it held out hope of expressing all of mathematics in the language of sets. At the time it seemed at least possible that Cantor's ideas would lead to the axiomatization of all of mathematics. The result, however, was more complicated.

The Russell Paradox

Cantor's set theory offered a new way of thinking about mathematics. Points, lines, planes, numbers, arithmetic operations, and spaces—it seemed that everything could be expressed in the language of sets. One could talk about a collection of functions or one could describe a single function in the language of sets. Every function began to be understood as a special type of relation between a domain (a set) and a range (another set). This approach seems natural and uncontroversial to mathematics students today, but there was initially great controversy about Cantor's theory of sets.

The beauty of Cantor's creation is that it held out the promise of a new way of perceiving mathematics. With Cantor's abstract formulation of mathematics, mathematicians could concentrate on the relations between sets. This made certain mathematical facts that had been difficult to understand suddenly transparent. Cantor's work pushed back the frontiers of mathematics. Suddenly mathematicians could approach problems from a highly abstract and general point of view. Instead of working with individual functions, for example, they could work with a whole class of functions simultaneously. With this kind of generality many mathematicians hoped that eventually it would be possible to discover and express the set-theoretic relations that lie at the foundation of mathematics. Cantor's discoveries, which were once perceived by many as abstract nonsense, describing nothing at all, now seemed to provide a language for describing everything. It was a tremendous success, and one that Cantor, toward the end of his life, must have recognized.

Cantor's last years were not easy, but despite periods of mental illness he continued to work on the theory of sets throughout his life. During this time he noticed a paradox associated with the theory of sets. This was a different kind of difficulty from the ones Cantor had faced before. Cantor's previous discoveries about sets had impressed many mathematicians as paradoxical, but the early definitions and theorems were not contradictory in a logical sense. The early theorems were sometimes counterintuitive, but Cantor's arguments had prevailed because they were logically correct. Cantor's newly discovered paradox, however, revealed a strange and fundamental shortcoming in his formulation of set theory. This discovery, unlike his initial work on sets, did not attract much attention. For whatever reasons, ill health or uncertainty about the meaning of his discovery, Cantor allowed the paradox to remain largely unnoticed. Soon, however, other mathematicians had detected the same logical error in Cantor's formulation. One of the first and certainly one of the loudest mathematicians to detect the problem was the British philosopher, mathematician, and social activist Bertrand Russell (1872–1970).

As a boy Bertrand Russell was privately tutored. Growing up, he had lots of time to think and excellent teachers from whom to learn. As a consequence he began to grapple with difficult philosophical problems when he was still quite young. One of the problems that fascinated him as a teenager was the problem of certainty. Russell was less interested in what people "knew" to be true than he was in how people had begun to know it. (As an adult he wrote a number of books on the limits of understanding; *An Inquiry into Meaning and Truth* is one such work.) Russell eventually enrolled in Trinity College, Cambridge, where he distinguished himself through outstanding scholarship. Despite his enthusiasm for academic work, these interests did not define him.

Russell's interests always extended beyond scholarly activity. He always enjoyed writing nonscholarly as well as scholarly works and clearly enjoyed creating controversy. He worked hard at both. He wrote many fine books and articles, some for the specialist and some for the lay reader. Russell also held strong pacifist views and during World War I he was fined and jailed for six months for his activities against the war. Russell was undeterred. He used his time in jail to write more about mathematics. Russell's most famous mathematical work, the three-volume *Principia Mathematica*, written with his former tutor, the mathematician Alfred North Whitehead,

Bertrand Russell preparing to address an antinuclear weapons rally at Trafalgar Square. In addition to his work as a mathematician and philosopher, Russell was a noted political activist. (Austin Underwood, McMaster University Libraries)

was an attempt to derive mathematics from first principles. This was an immense undertaking that neither author considered completely successful. It has been often said that though the book had two authors only one person has ever read the work cover to cover. In any case *Principia Mathematica* proved to be an influential text among logicians and mathematicians interested in the foundations of the subject. Russell continued to write about philosophy and mathematics throughout his life. He also remained active in social causes. In his last years he vociferously opposed the war in Vietnam.

With respect to his studies in set theory Russell took time to consider Cantor's ideas seriously. He was probably not the first to notice what has come to be known as the Russell paradox, but he almost certainly took more apparent delight in describing the paradox than anyone else. The Russell paradox is a beautiful example of the type of conceptual difficulty that Russell enjoyed uncovering. The paradox revealed something quite unexpected about Cantor's version of set theory at a time when set theory was being incorporated into all of mathematics. The paradox also illustrates how careful we must be, even when doing something as simple as describing a set, if we hope to prevent logical contradictions.

To understand the Russell paradox, we begin with the problem of defining a set. In our early school years, when we first learn about sets, we often just list the elements in a set of interest. Rather than writing a few sentences to specify the set consisting of all U.S. states beginning with the letter A, we may, simply list the states: Alaska, Alabama, Arizona, and Arkansas. Explicitly listing the elements in a set is fine for small sets, but for larger, more abstract sets, stating a condition that enables us to test whether any given object is a member of the set or not is wiser. The idea is simple enough: When presented with an object we simply compare the object against the condition of membership. If the object meets the condition of membership then it belongs to the set. If the object does not meet the condition of membership then it does not belong to the set. For example, suppose we let the letter S represent the set of all pencils. "The set of all pencils" is the condition that defines membership in the set. Every pencil belongs to this set; at the same time our condition of membership excludes all frying pans, automobiles, and any animal whose name begins with the letter a. This much is "obvious." What is not so obvious is that this simple method of defining sets can also lead to logical contradictions.

Notice that one consequence of defining the set S as the set of pencils is that *only* pencils belong to S. In particular the set S, which is a set *of* pencils, is not itself a pencil, and so the set S is *not* an element of itself. If, however, we define the set T as the set of all objects that are *not* pencils then not only do frying pans, automobiles, and animals with names beginning with the letter a belong to T, but T also belongs to itself. The reason is simple: T meets the condition of membership. It is an object that is not a pencil. We can only conclude that the set T belongs to itself.

Defining a set through a simple membership condition as Cantor had done can lead to logical contradictions. To see the problem, let *U* represent the set consisting of all sets that are *not* members of themselves. In particular *U* contains *S*, the set of all pencils, which is defined in the previous paragraph. It may appear that *U* has a reasonable membership condition until we ask whether or not *U* is a member of itself. Since the answer to this question should be either yes or no, we consider each possible answer in turn:

1. If *U* is *not* a member of itself, then *U* meets the criterion for membership (in *U*) and so *U* must be a member of itself. This is a contradiction, of course. *U* cannot be both a member of itself and not a member of itself, and so our first conclusion is that we err when we assume that *U* is not a member of itself.

2. Suppose *U is* a member of itself, and recall that *U* is defined as the set containing all sets that are not members of themselves: If *U* belongs to *U* then *U* is *not* a member of itself—again a contradiction.

These two contradictions taken together show that *U* never existed. This is Russell's paradox, and here is an easy way to remember it: In a small town with one restaurant and one chef, the chef cooks for all those who do not cook for themselves. This statement makes perfect sense for everyone but the chef.

Russell's paradox can be used to disprove the existence of the concept of a universal set. The *universal set,* a set that mathematicians had originally found quite convenient to use, is the "set of all sets." It is what we would get if we could form a gigantic set in which each element of the universal set is itself a set. The universal set, if it could exist, would contain *all* sets. *If* it existed, the universal set would contain the set of all pencils, the set of all frying pans, the set of all automobiles, and so on, but if the universal set existed, then it, too, would belong to itself, because it would meet the condition of membership: The universal set is a set. In fact the universal set would merely be one set among many that belong to the universal set, so the universal set would be a proper subset of itself. As a consequence we need to accept the existence of an even larger set that would contain all sets including the universal set, and therein lies the same contradiction. We *define* the universal set as the set that contains all sets. In particular there is no set larger than the universal set. Every set must belong to the universal set, but we have just shown that if the universal set existed then there must also exist an even larger set that contains it as a proper subset. The contradiction shows that we made an error in assuming the existence of the universal set. The only possible conclusion is that the universal set does not exist. No set can contain everything.

Resolving the Russell Paradox

The Russell paradox and other early set-theoretic paradoxes were associated with the existence of improperly defined sets. One approach to placing set theory on a firm mathematical basis was to express the basic ideas of the theory as a collection of axioms chosen in such a way as to rule out the existence of such sets.

Expressing a branch of mathematics through a collection of axioms was not a new idea. This was precisely the approach adopted by the Greek geometer Euclid more than 2,000 years before Cantor's birth. In his most famous work *Elements*, an introduction to geometry, Euclid began by listing a set of axioms, postulates, and definitions. These statements were intended to define the subject matter of the book. The axioms and postulates described the basic properties of the geometry. (Today mathematicians make no distinction between axioms and postulates, and we also use the words interchangeably from now on.) Once the axioms were listed, the act of discovery in the geometry of Euclid consisted of making nonobvious logical deductions from the axioms. These logical deductions are called theorems. In a sense the theorems are nothing new. The information contained in the theorems was present all of the time in the axioms. A mathematician's job is simply to reveal the information that the axioms contain.

Reducing discovery to the act of making logical deductions from a set of axioms is what makes mathematics different from science. In science when there are several competing theories, the "correct" theory is determined by experiment. If experimental results conflict with mathematical reasoning, it is the mathematical reasoning that is modified or abandoned, not the experiment. Mathematicians, on the other hand, care nothing for experiment. It is irrelevant to their subject. The goal of the mathematician is to establish an unbroken chain of logical arguments from one proposition to the next. As long as the mathematics is logically coherent, nothing else matters.

The importance of the axiomatic method was recognized early. Its importance becomes apparent when one tries to do mathematics without axioms. Early in the history of mathematics, before Euclid, Greek mathematicians realized that working without axioms was not at all satisfactory. They recognized that it was not enough logically to deduce condition C from condition B and condition B from condition A. The problem is that this chain of logical inferences just keeps extending further and further back. In our example everything in the chain depends upon the truth of A. If condition A is true then so are conditions B and C, but if condition A is false then we have learned nothing at all about either condition, B or C. The Greeks recognized that they were faced with an endless chain of implications, and their solution was to *begin* their deductions from a set of axioms whose truth was not open to question. These axioms cannot be proved within the confines of the subject because they define what the subject is. If we change the axioms we change the subject. The mathematical method is to begin with axioms and develop the subject from there.

Except for the geometry of Euclid not much thought was given to formal axiomatic systems until the 19th century, when several mathematicians, foremost among them the Russian Nikolai Ivanovich Lobachevsky (1793–1856), proposed a different set of axioms from the one adopted by Euclid. From these axioms Lobachevsky developed an exotic, new geometry. This geometry clashed with the commonsense notions that most people, even Lobachevsky, had about space, but Lobachevsky's geometry was sound in a mathematical sense, because it contained no internal contradictions. Mathematicians began to see the choice of one set of axioms over another as largely a matter of personal taste.

This is not to say that any set of statements qualifies as a set of axioms. Axioms must be free of contradictions in the sense that they cannot give rise to a statement that can be proved both true and false. They should be complete in the sense that those mathematical operations or results that mathematicians see as an integral part of their subject can be justified by the axioms. Finally, the axioms should be free of repetition in the sense that one cannot prove one axiom as a consequence of the others. At first Lobachevsky's work was ignored, but his geometry and other subsequent attempts to axiomatize various branches of mathematics eventually attracted a lot of attention from other mathematicians. They realized that it was Lobachevsky's highly abstract, axiomatic approach to mathematics that held out the promise of a mathematics free of logical errors. It was with this understanding of mathematics that the German mathematician Ernst Zermelo (1871–1953) turned his attention to Cantor's set theory and the problems posed by Russell's paradox.

Zermelo was born in Berlin into comfortable surroundings. He was educated at universities in Berlin, Halle, and Freiburg, and he received a Ph.D. in mathematics from the University of Berlin. Initially Zermelo studied hydrodynamics, the mathematical modeling of fluid flow, but he soon turned his attention to set theory. As Cantor and Russell had, Zermelo had discovered what is now known as Russell's paradox, and he decided that set theory could be saved by axiomatizing the subject in such a way as to rule out the existence of improperly defined sets. His solution was a set of seven axioms. From these seven axioms it was possible to deduce all of the major results of Cantor's set theory. The axioms were also chosen in such a way as to make deducing the paradoxes that had marred Cantor's original conception impossible. It was a great accomplishment.

One would think that Zermelo's accomplishment would have ended the controversy about the subject, but that was not the case. In order to accomplish his goal Zermelo needed an axiom that mathematicians everywhere now know as the axiom of choice. The *axiom of choice* is a statement about the existence of sets. It begins with a collection of sets, each of which is nonempty and disjoint from all the others. (A collection of sets is said to be disjoint when no two sets in the collection share a common element.) Essentially the axiom of choice states that given any collection of disjoint, non-empty sets, it is possible to create a new set that shares exactly one element in common with each of the sets in the original collection. It sounds innocent enough, but the axiom of choice was the cause of many objections. The source of the objections was the idea that the axiom applied to *any* collection of nonempty sets. Even if we were considering as many sets as there are real numbers, for example, the axiom of choice assures us of the possibility of choosing elements from each set. (Of course with so many sets the word *choose* has to be used loosely.) This selection process is something that we might imagine doing, but many mathematicians objected to the axiom because it was not "constructive." It was not always possible to devise an algorithm that would yield the necessary choices.

The axiom of choice allowed mathematicians to imagine the existence of an operation that had no basis in "reality," but without the axiom of choice many of the most interesting results of set theory could not be derived. Mathematicians were faced with a new kind of choice: They could abandon new and useful mathematical ideas because those ideas were based on nonconstructive procedures, or they could adopt the axiom of choice and accept a new kind of mathematics, a mathematics that consisted of ideas and solutions that were occasionally nonconstructive. Even today a few mathematicians object enough to the axiom of choice to refuse to use it. Other mathematicians use the axiom of choice when they have to but avoid it when they can. Others embrace it wholeheartedly.

Subsequent work showed that Zermelo's initial attempt to correct Cantor's theory of sets was not entirely successful, and Zermelo's work was slightly revised to incorporate ideas by the German-born Israeli mathematician Abraham Adolph Fraenkel (1891–1965), the Norwegian mathematician Thoralf Albert Skolem (1887–1963), and some later work of Zermelo himself. The resulting system of axioms is known as ZF set theory or sometimes ZFC set theory. (The "C" in ZFC is used to indicate that the axioms include the axiom of choice.) From the ZFC axioms one can derive the main set theoretic results discovered by Cantor, but the resulting theory avoids the paradoxes that marred Cantor's formulation, which is sometimes called the "unrestricted comprehension schema." Alternatives to the ZFC formulation of set theory have been proposed over the years with the goal of creating a theory of sets with certain additional characteristics. These alternatives have not been widely adopted. ZFC set theory continues to form the foundation for much of modern mathematics.

Zermelo made other important discoveries about set theory, but his career as a researcher was cut short by ill health. He had begun work at the University of Berlin and later worked at the University of Zurich. In 1916 he left Zurich as a result of poor health. He spent the next 10 years recovering from his illness. When he recuperated, he found a position at the University of Freiburg in 1926, but he resigned in 1935 in a protest over Nazi policies. At the end of the war, in 1946 he applied to the University of Freiburg to be reinstated in his old position, and his request was granted. He died seven years later.

Cantor's ideas enabled mathematicians to manipulate infinite sets successfully. This was extremely important to the development of all future mathematics, because so much of mathematics is about the infinite. Furthermore his very general language of sets enabled mathematicians to perceive their work in new and more productive ways. Set theory became the language of mathematics. Many of the best mathematicians of the day tried to incorporate Cantor's ideas of the infinite into a new and more rigorous approach to math. Their efforts have been only partly successful. Nevertheless Cantor's approach to mathematics remains at the heart of the subject to this day.

12

CANTOR'S LEGACY

In the early years of the 20th century there were many mathematicians interested in the foundations of mathematics. In addition to Ernst Zermelo, two of the most prominent names associated with the attempt to understand mathematics at its most fundamental level were Bertrand Russell, the author of Russell's paradox, and the German mathematician David Hilbert (1862–1943). Hilbert received a Ph.D. from the University of Königsberg. (Königsberg is now called Kaliningrad and is part of Russia. At the time it was an important German cultural center.) He remained at Königsberg for several years after graduation as a teacher. Eventually he was hired at the University of Göttingen, where he remained for the rest of his working life.

Hilbert was a prolific mathematician, who made important contributions to many areas of mathematics and physics. He helped to establish the branch of mathematics now called functional analysis, in which functions are studied in the context of set theory. He also helped found the theory of infinite-dimensional spaces. But Hilbert did more than find answers; he also famously posed questions. Today despite the many solutions that Hilbert found, he is best remembered for a series of questions that he challenged others to answer. In 1900 he posed 23 mathematical problems that he considered important to the future of the subject. Among these problems were questions that directly related to Cantor's work. Hilbert wanted to establish the truth of Cantor's continuum hypothesis, and he wanted to know whether it is possible to well-order any set. (We do not describe the concept of well-ordering here. We mention it because this question was quickly answered

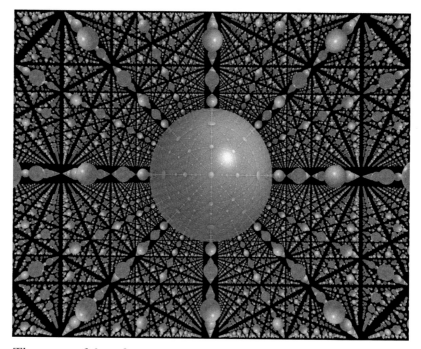

The concept of the infinite is one that continues to attract the attention of artists, mathematicians, and philosophers. (Casey Uhrig)

by Ernst Zermelo, and Zermelo's proof required the axiom of choice.)

Hilbert's personal prestige and his acumen in choosing the problems seized the attention of mathematicians around the world, and a great deal of research during the 20th century was devoted to solving these problems. Many—but not all—of the problems were solved, and as mathematicians worked to unravel Hilbert's problems a great deal of interesting and useful mathematics was discovered along the way. Hilbert's choice of problems affected the direction of mathematical research—even long after his death.

Hilbert had always been interested in the fundamental ideas and relations that underpin the subject to which he devoted his life. It had long been recognized, for example, that there are logical flaws in Euclid's formulation of his geometry. One of Hilbert's early successes was to revisit the geometry of Euclid and provide a set of 21 axioms that provide a complete and logical foundation for this most ancient of all geometries. After 2,000 years of error Hilbert was the first to present a completely rigorous foundation for Euclid's geometry. For Hilbert this was just the beginning. He had a larger goal: He wanted to do for the entire field of mathematics what he had done for Euclidean geometry. His method was, in a sense, mechanical. Hilbert wanted to strip away all extraneous meanings from mathematics. In geometry for example the meanings that we attach to *point*, *line*, and *plane* can help us understand both the motivation for studying geometry and any applications that arise out of the study of the subject, but these meanings are irrelevant to geometry as a branch of mathematics.

In geometry mathematicians are interested only in the relationships *between* points, lines, and planes; they are not interested in what these words mean. Hilbert's goal was to treat all branches of mathematics in a similar way. He wanted to eliminate extraneous meanings and retain only abstract sets of axioms. From these axioms the theorems of the mathematical system were deduced. It was a purely formal procedure. Hilbert's concept of formal mathematics had much in common with board games. In a board game we play by the rules. If we change the rules, we change the type of game that we are playing. In a board game our goal is to follow the rules, not create them. In mathematics once the axioms have been established all that remains is to deduce new theorems. The goal of every branch of mathematics is to use the axioms to prove new theorems. Mathematicians seldom create new axioms.

It may seem that once Hilbert had stripped all of the non-mathematical meaning from a mathematical system there would be nothing useful or interesting remaining. This is not true. Hilbert's formalized mathematical systems were like machines with the housings cut away to reveal the moving parts. We cut away the housing of a machine to see how it works, not to learn what the machine does. We can see what the machine does from the outside. To see *how* it does it, however, we need to look deeper. This was Hilbert's goal. No longer was there any question about whether the mathematical system under consideration was "true" or "correct" in the sense that it agreed with reality. Instead Hilbert's formalizations allowed us to see how—or even whether—a mathematical system generates good, logically consistent mathematics. Hilbert's goal was to look *inside* mathematics to discover the way it worked.

Keep in mind that in Hilbert's conception of formal mathematics the only question of interest is whether the theorems are direct, logical consequences of the axioms. This is a question best answered by examining the formal, abstract axioms that defined the subject. In this very rarefied atmosphere, the questions of the completeness and consistency of the axioms become central considerations. A set of axioms is *complete* if, given a statement in the mathematical system described by the axioms, the statement can always be proved either true or false. A complete set of axioms rules out the possibility of unprovable statements. A set of axioms is consistent provided every statement that is a consequence of the axioms is either true or false but not both true and false.

Hilbert began to search for criteria that would enable him to determine the completeness and consistency of particular sets of axioms. Hilbert, to be sure, never expected an easy answer to the questions of completeness and consistency. The fact that one has not yet found an unprovable statement does not mean that the axioms are complete. Nor does the fact that one has yet to find a statement that is both true and false mean that no inconsistency

$*54\cdot43.$ $\vdash :. \alpha, \beta \,\epsilon\, 1 . \supset : \alpha \cap \beta = \Lambda . \equiv . \alpha \cup \beta \,\epsilon\, 2$

Dem.

$\vdash . *54\cdot26 . \supset \vdash :. \alpha = \iota'x . \beta = \iota'y . \supset : \alpha \cup \beta \,\epsilon\, 2 . \equiv . x \neq y .$

[$*51\cdot231$] $\equiv . \iota'x \cap \iota'y = \Lambda .$

[$*13\cdot12$] $\equiv . \alpha \cap \beta = \Lambda$ (1)

$\vdash . (1) . *11\cdot11\cdot35 . \supset$

   $\vdash :. (\exists x, y) . \alpha = \iota'x . \beta = \iota'y . \supset : \alpha \cup \beta \,\epsilon\, 2 . \equiv . \alpha \cap \beta = \Lambda$ (2)

$\vdash . (2) . *11\cdot54 . *52\cdot1 . \supset \vdash .$ Prop

From this proposition it will follow, when arithmetical addition has been defined, that $1 + 1 = 2$.

A theorem from Whitehead and Russell's Principia Mathematica. *In their zeal to create a completely unambiguous form of expression, they authored a book that has always been more widely admired than widely read.* (University of Michigan Historical Math Collection)

exists. Hilbert worked hard on the problems of completeness and consistency of sets of axioms. His plan of formalizing all of mathematics was not widely emulated, however.

Bertrand Russell took a different approach to understanding the foundations of mathematics. He wanted to show that pure mathematics is simply a branch of logic. Logic is that branch of knowledge that is concerned with the laws of rational thought. By beginning from first principles Russell hoped to derive mathematics from a sequence of carefully ordered logical steps and in this way prevent the logical paradoxes that had attracted his attention and the attention of others. This was one goal of the three-volume set *Principia Mathematica* that Russell and his former teacher at Cambridge Alfred North Whitehead wrote together.

In *Principia* Russell and Whitehead hoped to recast mathematics. For them mathematics was not really about numbers; it was an exercise in logical thinking. Russell famously expressed his attitude about the whole field when he asserted that in mathematics one does not know what one is talking about nor whether what one is saying is true. Content and truth, Russell enjoyed asserting, are not the province of the mathematician. All that matters is whether the theorems are logically deduced from the axioms, and whether the axioms themselves are logically consistent with one another. Russell's paradox, for example, revealed a logical error. It has little to do with numbers. Because it represented a logical failure, one could correct it only by sharpening one's logical tools. Russell and Whitehead spent years developing the necessary ideas and notation to express their ideas. They were not the first to attempt this, and they were heavily influenced by those who preceded them, but Russell and Whitehead were the most successful mathematicians of their time in expressing mathematical ideas in the language of logic.

Russell and Whitehead were in their own way attempting to create a formal system that would display the workings of mathematics in a purely abstract setting. Over the course of their studies Russell and Whitehead returned to the set theory of Cantor and were able to deduce some of Cantor's main ideas by using their approach. They returned to the arithmetic of ordinary finite

THE PEANO AXIOMS

During the late 19th and early 20th centuries, many of the most prominent mathematicians devoted a great deal of effort to formalizing mathematics. One of the most influential was the Italian mathematician and logician Giuseppe Peano (1858–1932). Peano had a special interest in the logical foundations of mathematics and hoped to formalize all of mathematics. His best-known effort in this regard is called the Peano axioms, which he developed in conjunction with others, and which can be used as the logical foundation for arithmetic and that branch of mathematics called number theory. Here are Peano's five axioms:

(The letter N denotes the natural numbers.)

1.) 1 is a member of the set N. (This shows that the set N is not empty.)

2.) If n belongs to the set N, then $n + 1$ belongs to N. *(n + 1 is called the "successor" to n.)*

3.) 1 is not the successor of any element in N.

4.) If $n + 1 = m + 1$ then $m = n$.

5.) Let S be any subset of N such that S contains 1. Suppose that the set S has the additional property that if S contains n then S also contains $n + 1$, then $S = N$. (Axiom 5 forms the basis for the method of proof called mathematical induction.)

It was Peano's hope that these few sentences and an accompanying set of statements defining the concept of equality would form a complete and consistent set of axioms for arithmetic. Russell and Whitehead further extended Peano's results in their work *Principia Mathematica,* and Peano's axioms are widely quoted even today, a century after they were first proposed.

But today we know that completeness cannot be achieved by any set of axioms. And with respect to consistency, the best that can be said is that no inconsistencies have been found yet. (See the Afterward for an interview with Professor Karlis Podnieks.) It is, nevertheless, remarkable how enduring and useful Peano's simple-looking set of axioms have proven to be.

numbers and transfinite numbers, and they likewise reinterpreted these ideas in the language of logic. They did a lot, but there was a limit to their approach past which they could not go. Their reach exceeded their grasp. After three large volumes Whitehead and Russell stopped. By their own admission their attempt to exhibit even basic mathematics as a consistent formal system had not been entirely successful.

Kurt Gödel and the Axiomatic Method

The efforts of Cantor, Russell, Whitehead, Hilbert, and Zermelo to formalize mathematics was derailed in a surprising way by discoveries made by the Austrian-born mathematician Kurt Gödel (1906–78). Gödel identified limitations to the axiomatic method, thereby proving that there were limitations to what one could learn from the mathematical method that had been in use since Euclid and that continues to be used today.

After receiving his Ph.D. from the University of Vienna in 1929, Gödel remained at the university as a faculty member. During this time, there was a great deal of political instability in Europe. The Nazi Party had come to power in Germany, and in Austria there was much sympathy for the ideas of National Socialism. Gödel was not especially political, but he was profoundly affected when, in the late 1930s, a close friend and colleague of his was murdered by a student who was a Nazi. In 1938, Gödel moved to the United States, where he joined the Princeton Institute for Advanced Study. He remained there for the rest of his working life.

What makes Gödel so important to the history of mathematics is his 1931 discovery that the efforts of Hilbert and others to formalize all of mathematics were naive. He showed that certain assumptions that were then common about the nature of mathematics were false.

Keep in mind that mathematicians since Euclid had understood mathematics as the art and science of deducing theorems from the axioms, undefined terms, and definitions that comprised the subject. They believed that once the axioms and requisite vocabulary had been specified, the entire subject was determined. (This is still the case.) But earlier generations of mathematicians took these statements to mean that *all* statements about the subject, whether true or false, were but logical consequences of the axioms. Gödel deduced something quite different. He showed that there were certain statements that could be made about any complex mathematical system—in Gödel's case, arithmetic—that were not logical consequences of the axioms. At first glance, it may seem that each statement must be true or false, but the situation is more complex. The meta-mathematical statements that Gödel was concerned with represent unsolvable problems in the sense that they cannot be proved or disproved from within the system defined by the axioms—that is, he proved the existence of mathematical sounding statements that are, logically, neither true nor false.

Gödel's method of proof demonstrates that such statements must exist. He does not, however, provide a ready means of testing whether any particular statement is a logical consequence of a given set of axioms. Instead, his discoveries, summarized in his incompleteness theorems, demonstrate that Hilbert's intuitive belief that it is possible to create a system of axioms that is logically complete and consistent was incorrect. There are statements about any complex mathematical system the truth of which cannot be deduced from the axioms that define the system. These statements lie "outside" the system.

Gödel demonstrated that statements about a mathematical system—e.g., arithmetic, Euclidean geometry, and probability theory—fall into one of two types. Some statements can be shown to be logical consequences of the axioms. These statements are intrinsic to the subject. In other words, they can be proved either true or false. They are consequences of the axioms, definitions, and undefined terms that constitute the subject. These statements are what constitute mathematics as mathematics has been understood for thousands of years. But there are other statements, sometimes called meta-mathematical statements, about any mathematical system that do not lie within the logical framework determined by the axioms, definitions, and undefined terms that constitute the subject. (This is often shortened by saying that these statements cannot be deduced from the axioms. We will follow that practice here.) These statements can be said to lie outside the subject determined by the axioms. The truth of such meta-mathematical statements cannot be addressed from within the subject that they purportedly describe.

Goldbach's conjecture is sometimes mentioned as a possible example of a statement about arithmetic that may not be a logical consequence of the axioms that define that subject. In 1742 the Russian mathematician Christian Goldbach (1690–1764) hypothesized that every even integer greater than 2 can be written as the sum of two prime numbers. Here are four examples that demonstrate the conjecture (for the first example, recall that 2 is a prime number): $4 = 2 + 2$; $12 = 7 + 5$; $30 = 23 + 7$; and $100 = 97 + 3$. The number on the left side of each equation is even, and the two numbers on the right side of each equation are prime. Since Goldbach's time, many other even integers have been checked, and in each case mathematicians have found two prime numbers which, when added together, equal the even number in question. But this proves nothing. There are infinitely many even numbers, and most of them are too big to write down, or imagine, or process using the world's fastest computers, and this will always be true. Consequently, proving that a comparative handful of even numbers satisfy Goldbach's conjecture fails to prove the nonexistence of an even number greater than 2 that cannot be written as the sum of two primes.

To many people, Goldbach's simple-sounding statement seems as if it must either be true or false: Either every even integer can be written as the sum of two primes or not. But despite a great deal of work on the part of many mathematicians, only partial results have been obtained. It has, for example, been shown that any sufficiently large even integer is the sum of no more than four primes, which is an interesting result but one that falls short of proving Goldbach's conjecture.

After so much effort, it is natural to wonder why Goldbach's conjecture has been so resistant to proof. It could be, of course, that the conjecture is simply difficult (as opposed to impossible) to prove or disprove, and if this is the case someone may eventually find a way to either prove or disprove it. On the other hand, Goldbach's conjecture may be an example of a statement about arithmetic that is not a logical consequence of the axioms that define the subject. If it is not a logical consequence of the axioms that define arithmetic, it cannot be deduced from those axioms. But if Goldbach's conjecture is not a consequence of the axioms, a third question arises: What sort of meaning can be attached to such a statement? (Even if it turns out to be the case that Goldbach's conjecture cannot be proved true or false using the axioms of arithmetic, some mathematicians and philosophers would still be willing to consider whether or not the conjecture is true in some expanded system of axioms.)

One may suppose that what Gödel actually did was to identify a flaw within the set of axioms defining arithmetic. It may seem that he discovered that the particular set of axioms that he considered—axioms that he obtained from Russell and Whitehead's *Principia Mathematica*—were simply incomplete, and that the solution to the problem of incompleteness that he identified is to create a better, more complete set of axioms, but this interpretation is not quite correct. The difficulty Gödel uncovered cannot be fixed. He showed that there is no complete set of axioms for arithmetic. Gödel's discovery places a limitation on the axiomatic method itself. If we identify mathematics as the axiomatic method, then Gödel's work places limits on what can be discovered using mathematics.

To be clear, suppose that it is proved that some statement—Goldbach's conjecture, for example—cannot be deduced from a particular set of axioms. One could, presumably, fix the problem by introducing one or more additional axioms that would make the statement either provable or disprovable. Suppose that from this new, augmented set of axioms, definitions, and undefined terms one could deduce all the familiar results of the "old" arithmetic and in addition prove or disprove the statement in question. Gödel's proof does not rule out this possibility. Instead, what Gödel showed is that *no matter how the axioms are chosen* there will always exist some statements about the system the truth of which cannot be deduced from the axioms. In this sense, the situation is hopeless. Complete sets of axioms for mathematical systems, the kinds of sets of axioms that Hilbert sought to discover, cannot be discovered because they do not exist.

Alan Turing and His Machine

The British mathematician, logician, and computer scientist Alan Turing (1912–54) sought to extend the results of Kurt Gödel in a useful and creative way. Recall that Gödel had showed that there are certain statements that one can make about a mathematical system—in the case considered by Gödel the system was arithmetic—that cannot be proved true or false using only the axioms, definition, and undefined terms of the system itself. Although Gödel showed that such statements exist, he did not provide a general rule for identifying them. Turing examined the problem from a different point of view.

Turing had a particularly creative and adaptive mind and a fondness for interdisciplinary problems, those problems that span what many people consider separate branches of knowledge. Turing contributed to mathematics, computer science, cryptology, and morphogenesis, the study of how form and structure develop in living organisms. He made two contributions that are especially important for the history of numbers: He extended Gödel's results on incompleteness, and his work contributed to the discovery of a new type of number.

With respect to Gödel's incompleteness theorems, Turing sought to imagine a machine that could determine whether a statement about a mathematical system was "undecidable," which is the technical term for a statement about a mathematical system the truth of which cannot be deduced from the axioms that define the system. This machine, now called a Turing machine, is a mathematical abstraction that illustrates the idea of an algorithm.

An algorithm is defined as a step-by-step procedure for solving a problem or a class of problems. In particular, an algorithm must be mechanistic in the sense that it leaves no room for individual judgment. It must also be practical in the sense that it can be written down. One consequence of this last condition is that the algorithm must be expressed in finitely many steps. (Infinitely long algorithms cannot be written down.) The algorithm may, however, be recursive in the sense that a particular sequence of steps within the algorithm can be repeated as often as necessary in order to obtain a particular result. While there is no hard limit to how often such a subprogram can be repeated in the course of solving a problem, the algorithm of which it is a part consists of finitely many lines of "code." It is, therefore, finitely long.

The Turing machine is a device that can assume a finite number of states. Its input and output consists of a tape divided into squares. The tape can be imagined as infinitely long, and all but finitely many of the squares are left blank. (That only finitely many squares contain symbols expresses the idea that the algorithm that determines the machine's operation must be finitely long.) Each square that is not blank contains a single symbol drawn from an alphabet consisting of finitely many symbols. The machine has a read-write "head" that reads one symbol at a time. Once the symbol is read, the machine responds. It may (1) erase the symbol and leave the square blank, (2) erase the symbol and print a new symbol on the square, (3) leave the square unaltered, and/or (4) advance the tape forward or backward, where the terms "forward" and "backward" are terms of convenience; there is no preferred direction. Finally, (5) the machine may come to a stop. The action that the machine takes is determined by the square that it scans and the state of the machine as it scans the square.

To understand what is meant by the "state of the machine," think of the machine as consisting of a large collection of on-off switches. The state is determined by the configuration of the switches—that is, the pattern formed by the off and on switches. When the Turing machine finishes its work—or more precisely, *if* the Turing machine finishes its work—it stops. The output of the machine is represented by the completed tape. Turing's conceptual machine operates without any time limitation. It accepts input, manipulates it, and produces output, and this is what makes the Turing machine a sort of proto-computer. Its value lies in the way it represents the abstract idea of an algorithm.

Today, most people regard a computer as a device for interacting with other computers through the Internet, or as a device for playing music files, or as a device to control other devices such as printers. For Turing, the purpose of his machine was to determine whether or not a particular proposition about a mathematical system is decidable—that is, whether or not the statement could

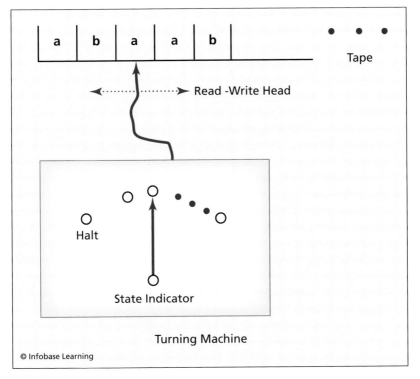

Diagram of a Turning machine. The dial indicates the state of the machine.

be deduced from the axioms. He phrased the problem in such a way that if his machine came to a halt then the proposition under consideration was decidable. If, on the other hand, the machine did not come to a halt then the proposition was not decidable. It is, in effect, a "decidability tester."

But there is a difficulty: The fact that a machine that could, in principle, run forever has not stopped in five minutes, five days, or 5,000 years does not mean that it might not stop at some later date provided that the machine is allowed to operate a little longer. The fact that such a machine has not stopped, therefore, does not give any information about whether it might not eventually stop. Consequently, if a machine is programmed to determine whether or not a particular proposition is decidable, and if it has not stopped at the end of some finite time, then it has revealed nothing about the decidability of the proposition under consideration. Running

FORMAL LANGUAGES TODAY

The formalized mathematics of Hilbert and others never did catch on among mathematicians. This is due, in part, to the difficulty of learning the necessary notation and, in part, to the fact that they offer only a modest increase in clarity. It is also true that Gödel's incompleteness theorem demonstrates that the goal of the formal approach cannot, in any case, be completely realized. In retrospect it may appear that the highly formalized symbolic language pioneered in the work of Cantor, Hilbert, Russell, Whitehead, and others was a dead end, and mathematically this may prove to be the case. (We should, however, keep in mind that in mathematics the last chapter is never written.)

There are many others, however, who believe that formal languages are now some of the most important languages in our culture. Like the formal languages envisioned by Hilbert, these formal languages are not *about* anything. They have no meaning in themselves. They simply give us a method of describing relations that exist between unknown classes of objects. These formal languages are computer-programming languages, and they give us a way to describe relationships that exist between variables. The languages are a set of rules to manipulate variables. We may or may not know—or care—what the variables represent. The computer performs the manipulations without giving any consideration to whether the outcome is "true" or "correct," and it displays results without regard to whether or not they agree with reality. The computer allows us to determine *only* what are direct, logical consequences of our assumptions, because the computer can only follow the rules that are encoded in the language. In this sense formal languages are now more important than ever. Some computer programs even resemble material in *Principia Mathematica* because of the abstract appearance of the symbols and the way that the language is expressed.

a Turing machine might, therefore, be an enormous waste of time. A better strategy would be to program one Turing machine—call it Turing machine A—to calculate whether a proposition is decidable, and then to use a second Turing machine—call it Turing machine B—to predict whether or not Turing machine A will halt. The problem can then be resolved by machine B. If (using machine B) one could predict beforehand that machine A would stop, then one would know that the proposition that machine A was evaluating was decidable. If, on the other hand, one could predict (again using machine B) that machine A would not stop no matter how long it ran, then one would know that the proposition under consideration was not decidable. Turing discovered that there does not exist an algorithm (as represented by Turing machine B) that could predict beforehand and under general conditions whether or not Turing machine A would halt. Because one cannot, even in theory, predict the behavior of one Turing machine with another Turing machine (and because in general one can learn nothing from watching the first Turing machine), one cannot determine whether or not a particular proposition is decidable. Because the concept of a Turing machine is flexible enough to represent any algorithm, Turing concluded that it is not possible to find a general procedure to identify nondecidable propositions. Called the halting problem, Turing's discovery reveals a further limitation of the axiomatic method.

A New Type of Number

Soon after his brief investigations into the nature of computation, Turing directed his considerable energies toward other types of problems. His research had already yielded a number of new ideas about the nature of algorithms, and these ideas were further refined by his successors. One consequence of this additional research is that some mathematicians came to perceive the work of Dedekind in a new light. Recall from chapter 8 that Dedekind had established a 1-1 correspondence between the set of all points on a line and the set of all real numbers—each set forming a continuum. Dedekind's work is often described as contributing to the arithmetization of analysis, by which is meant that he took the idea of continuity, which had previously been expressed using ideas that were geometrically plausible but not mathematically rigorous and made it rigorous. Dedekind's discovery came at a critical moment in the history of mathematics because the old ideas about number and the continuity of the number line were not rigorous enough to support the development of the new mathematics of his time. Mathematicians had discovered a number of paradoxes that could only be resolved with clearer ideas about the nature of numbers. By expressing his ideas about numbers in much more rigorous terms, Dedekind contributed to mathematical progress.

It took many years for mathematicians to fully appreciate what Dedekind's view of the real numbers means. In particular, embedded within Dedekind's model of the real number system is a 1-1 correspondence between real numbers and certain sequences of rational numbers. Sequences are, as a matter of definition, countable collections of numbers, which is another way of saying that every sequence can be placed in 1-1 correspondence with the set of natural numbers. The correspondence is sometimes expressed using function notation: The nth term of a sequence is written as $f(n)$, and sometimes the sequence itself is called $f(n)$. The assertion that every real number is the limit of a sequence of rational numbers can, therefore, be expressed as an assertion about the existence of a certain set of functions, which, for ease of reference, we will call set F. Every element of F is a function, but not every function belongs to F. A function, $f(n)$, belongs to the set F provided it satisfies three criteria:

1.) The domain of $f(n)$ is the set of natural numbers.

2.) The range of $f(n)$ is a subset of the set of rational numbers.

3.) The sequence $f(n)$ is a convergent sequence.

With respect to item 3, recall that a sequence of real numbers is said to be convergent provided there exists a point L on the real number line such that if we draw any circle about L—no matter how small—all but finitely many values $f(n)$ will lie within this circle. The sequence $f(n)$ is said to converge to L. To put it another way: The set of values $f(n)$ cluster about L is such a way that for all large values of n, $f(n)$ is "close" to L, where the radius of the circle that we draw about L conveys our definition of what we mean by close. The number L may or may not be rational, but if L exists, it must be unique. (To see why L must be unique, imagine that there were two distinct points, L and M, and that the sequence $f(n)$ clusters about L. To show that the sequence cannot also cluster about

M, draw two circles, one centered about L and the other centered about M and draw them so small that they do not overlap. Since, as a matter of definition, all but finitely many values of $f(n)$ lie within the circle centered at L, the circle about M must be almost empty in the sense that at most finitely many of the numbers $f(n)$ will lie within it. Consequently, if the sequence $f(n)$ converges to L, it cannot also converge to M.)

It is one thing to assert the existence of sequences $f(n)$ in F, each of which converges to a real number, it is another thing to try to compute them. Not all of the sequences in F are computable. As a matter of definition, a sequence is computable if an algorithm for the sequence can be executed on a Turing machine. In other words, in order that a sequence be computable there must exist a finite sequence of instructions—these are represented by the symbols that appear in the squares of the tape that runs through the Turing machine—such that using this set of instructions, the Turing machine can produce the sequence. The instruction set must be finite in length and written using an alphabet that consists of finitely many symbols.

But it is an immediate consequence of Georg Cantor's research that there are only countably many such instruction sets. And as Cantor well knew, there are uncountably many real numbers. (See, for example, *There Are More Real Numbers Than Rational Numbers* in chapter 11.) These "countably many such instructions sets" are exactly the *computable* sequences that belong to F. Since there is a correspondence between each real number and the sequences of rational numbers that converge to it, we can only conclude that many of the sequences in F are not computable. Those real numbers that cannot be placed in correspondence with a computable sequence in F are called uncomputable, where the term *uncomputable* means that they cannot be computed with a Turing machine. Because the set of real numbers is uncountable, and because the set of computable real numbers is countable, most numbers are uncomputable.

Despite the fact that there are more uncomputable than computable numbers, it is much easier to give examples of computable numbers. All rational numbers, for example, are computable. The division algorithm that all students learn in grade school can, for example, be executed on a Turing machine, and it is, in theory, adequate for expressing any rational number in decimal notation. Algebraic numbers, which are defined as those numbers that appear as roots of algebraic equations, are also computable. Some transcendental numbers—π is but one example—are computable as well. For practical purposes, then, there are enough computable numbers to satisfy the needs of any engineer or scientist. By contrast, very few examples of uncomputable numbers are known. The American mathematician Gregory Chaitin (b. 1947) produced the first example of an uncomputable number, a number that he calls Ω (omega). The number Ω can be defined, but it cannot be computed in the sense previously defined.

So far, we have defined uncomputable numbers by what they are not; they are not computable. In some ways, the situation is similar to that of the irrational numbers before Dedekind. For a very long time, irrational numbers were defined by what they are not; they are not rational. And as with irrational numbers, defining something in terms of what it is not, fails to provide much insight into what it is. What, then, is a more positive description of an uncomputable number?

To provide a more informative description of a noncomputable number, we can (again) use a Turing machine. Imagine that every number that we use in our Turing machine is expressed in base 2— that is, every number with which we will be concerned will appear as a sequence of zeros and ones. (This is not a restriction on our choice of number because every number can be expressed in base 2.) Choose one such number. Now imagine creating a computer program to write that number. At some level every computer program can also be expressed as a sequence of zeros and ones. (In fact, in the early days of computer programming, this was the only way that computer programs could be expressed.) We can now compare the length of any number (written in binary code) with the length of the computer program needed to write that number when that computer program is also expressed in binary code.

As mentioned earlier in this section, those numbers that are computable can be represented by computer programs that terminate, which in this situation means that the string of zeros and ones that represents the computer program is finitely long. Such programs can always be executed by a Turing machine. But for a noncomputable number—and for Chaitin's Ω in particular—the situation is a little more complicated. In order to compute the first n binary digits of an uncomputable number, one would need a computer program that is (roughly) n binary digits long. The more digits of an uncomputable number that we wanted to compute, the longer our program would need to be. Not surprisingly, one could write a program that uses more than n binary digits, but it is not possible to use a program that uses less than n binary digits. Because it is not possible to create a program of a fixed (finite) length to compute arbitrarily many digits of an uncomputable number, the digits that comprise an uncomputable number cannot be determined using a Turing machine.

There is still another interpretation. Embedded within the idea of uncomputability is the idea of randomness. One way of understanding a random sequence of numbers is as a sequence that cannot be generated by a Turing machine. To see why this is so, imagine that we are given a particular (infinite) sequence of numbers. If it is possible to write a finite set of instructions for a Turing machine that enables it to generate the sequence, then we have successfully "predicted" the sequence. The sequence cannot, therefore, be random since random sequences are, as a matter of definition, unpredictable. In this view, the defining characteristic of a random sequence is that there is no finite instruction set that would enable a Turing machine to generate the sequence. Because a random sequence cannot be generated using a finite instruction set, the only way to generate the sequence is to list the numbers that comprise the sequence. Any program that generates a random sequence must, therefore, be at least as long as the sequence itself, and this is exactly the situation with uncomputable numbers. Consequently, the existence of uncomputable numbers implies the existence of a random element in mathematics. That, at least, is the view of some.

There is no broad consensus among mathematicians about what the work of Gödel, Turing, and the discovery of uncomputable numbers means to mathematics. On one hand, Gödel and Turing's ideas reveal limitations on what can be learned by using the axiomatic method. Some mathematicians claim that the limitations are severe and that modern mathematics consists of a small number of provable statements awash amid a sea of undecidable, essentially random, metamathematical statements. Others claim that the incompleteness theorems help to identify the boundaries of mathematics. Those who believe mathematics consists largely of unprovable statements tend to see mathematics as more of an experimental discipline than as a purely deductive form of reasoning. On the other hand, most professional mathematicians continue to be concerned exclusively with deducing theorems from axioms. One cannot tell from listening to most contemporary lectures on mathematics that Gödel and Turing ever existed. It is not a question of whether Gödel and Turing are right. They proved that their results are correct. Nor is it a question of whether Chaitin's Ω is uncomputable. This, too, has been proved. The question is, instead, what do these results imply about the nature of mathematics?

There are certainly some conjectures that have long resisted proof, and it is not clear why. Goldbach's conjecture was mentioned earlier, but there are many more. How many of these are unprovable, and how many are just difficult to prove? To be sure, there are statements that resisted proof for centuries only to be proved in a nonobvious way using recently developed mathematics. Fermat's Last Theorem, a famous result in number theory, is an example of such a statement. Prior to the proof, there were some who questioned whether the statement could be proved at all. Now everyone agrees that it has been proved.

It will take time to determine what these discoveries about numbers, computability, and decidability mean, and future discoveries can (and probably will) change our perceptions. The fundamental theorem of algebra, for example, which states that every algebraic equation has a root, was considered to be a profound insight when the first proof of the result was produced two centuries ago, but today most mathematicians in the field of algebra would point to other more recent discoveries that they consider far more fundamental to algebra than the fundamental theorem. Time will tell whether the work of Gödel, Turing, and Chaitin changes everything, as Chaitin himself tends to claim, or whether these results are interesting insights into the limitations of the axiomatic method that leave mathematical research and our understanding of mathematics largely unchanged.

CONCLUSION

Numbers have attracted the attention of the curious throughout history. There is good reason to believe that interest in numbers predates recorded history. Some of the earliest historical records show that mathematicians and philosophers were already hard at work seeking to understand the nature of numbers and their uses. Throughout the intervening millennia, mathematicians of many different cultures and in many parts of the world have further expanded and refined the definition of number. This has been time well spent. The field of numbers has proven to be one of the richest of all human intellectual pursuits with some of the most important practical applications.

Progress has seldom been rapid. Questions about numbers and their uses have sometimes been difficult to solve, and often progress was slowed by the inability of mathematicians and others to accept results that were already known to them. Because numbers are so central to our consciousness, we are, whether we know it or not, emotionally as well as intellectually invested in them. This has always been true. The ancient Greeks, for example, regarded the discovery of irrational numbers as extremely significant, but Diophantus, one of the most important Greek mathematicians of his time, was reluctant to accept irrational numbers as answers to the problems he solved. Similarly, during the Renaissance mathematicians knew about complex numbers—they even used them during the process of computing the roots of polynomial equations—but they refused to accept imaginary numbers as final answers just as Diophantus had refused to accept irrational numbers as answers to the problems with which he was concerned. It was not until early in the 19th century that European mathematicians fully accepted the idea of negative numbers, although negative numbers had been known in Europe for centuries.

One more example from the 19th century: The prominent German mathematician Leopold Kronecker (1823–91) is perhaps best remembered for his remark, "God made the integers. All else is the work of man." The statement succinctly captures the view of Kronecker and many of his contemporaries about the nature of numbers. Less appreciated is Kronecker's determined and successful opposition to the efforts of Georg Cantor to obtain a position at the University of Berlin. Cantor was certainly one of the most creative and important mathematicians of all time, and he very much wanted to work at the University of Berlin, where Kronecker already worked. To be sure, Kronecker was a successful and in some ways an insightful mathematician, but their relative contributions were hardly comparable. This was apparent even while both men were alive, but Kronecker prevailed. His resistance to Cantor's appointment arose from his sometimes bitter objections to Cantor's discoveries regarding the nature of infinity and his development of a system of transfinite numbers. Kronecker understood Cantor's discoveries, but he could not accept them.

Today, mathematicians continue to develop new ideas about numbers. Some claim that the discovery of uncomputable numbers, for example, reveals the existence of a vast and previously unexpected mathematical landscape, one where the role of mathematical proof is marginalized and mathematics becomes more of an inductive than a deductive science. The discovery of the uncomputable marks, for them, the beginning of a revolution in mathematics. Of course, just saying that a revolution in mathematics has occurred does not make it so, and just as new and important mathematical ideas have taken generations to become fully incorporated into mathematics, there are many instances of inflated claims about the value of new ideas and their place in mathematics. The implications of more recent work about the nature of mathematics and the nature of numbers may not be apparent for generations. Our view of numbers and their meaning continues to evolve.

AFTERWORD

THE NATURE OF MATHEMATICS — AN INTERVIEW WITH PROFESSOR KARLIS PODNIEKS

Dr. Karlis Podnieks studied mathematics at the University of Latvia from 1966–71. In 1979 he received his Ph.D. in mathematics from the Computing Centre of the USSR Academy of Sciences in Moscow, where he did research on machine learning algorithms. Initially, his

research interests included work on the foundations of mathematics. In 1980 he turned his attention to the theoretical and practical aspects of computation, including computer programming, database design, information systems, and graphical tool development for business process modeling. He has, however, continued to write about the philosophical foundations of mathematics, and his writings are notable for their insight, humor, and originality. He became professor of information technologies at the University of Latvia in 2005. This interview took place in April 2009.

Karlis Podnieks (Courtesy of Karlis Podnieks)

J. T. Professor Podnieks, some mathematicians talk about "discovering" mathematics. Others talk about "creating" mathematics. The difference is important. One discovers things that have an existence of their own. Most people would agree that one discovers planets, for example, or species of birds. By contrast, one creates things that did not exist prior to the act of creation—symphonies, for example, or automobiles. Do you think that mathematics is created or discovered? Or to put it another way: How much of mathematics has an objective existence and how much do we simply imagine into existence?

K. P. I would prefer the term "inventing" instead of "creating." Speaking strictly, one discovers neither planets nor species of birds. One is inventing models of the world—or of parts of it. Some time ago, planets were thought of as lights attached to crystal spheres. Were these spheres "discovered" or invented? Some time later, a new model was invented in which planets were thought of as massive bodies orbiting the Sun. This picture remains stable after essential refinements of the model due to Kepler, Newton, Einstein, et al., and after the new evidence obtained recently by Gagarin, Apollo crews, etc. In fact, this stable part of model evolution is what people are calling "discovered final truth." As to symphonies and automobiles—after creation, they can be discovered just as planets were. Staying with the usual naive notion of "discovering truth about reality," we will never be able to understand the nature of mathematics.

J. T. I thought of using the word "invent" rather than "create," but I think that some decisions about mathematics—the choice of axioms for a particular discipline, for example, and the choice of problems to study—depend on aesthetics. Mathematicians sometimes make these choices because the results that they obtain appeal to their sense of beauty. In this sense, mathematics has a good deal in common with art. But I can see that some mathematics may better be described as "invented" because it is developed in response to specific, often predetermined problems—especially those arising in engineering and the sciences. Anyway, that was my thinking . . .

K. P.   On the above human "modeling panorama," where should the place of mathematics be? Physicists, chemists, biologists, economists, psychologists, et al. are inventing models for their "part of the world." Then, what are mathematicians doing?

For many years, I have been promoting the broadest possible notion of mathematical models. Many people think that mathematical models are built using well-known "mathematical things" such as numbers and geometry. But since the 19th century, mathematicians have investigated various nonnumerical and nongeometrical structures: groups, fields, sets, graphs, algorithms, categories, etc. What could be the most general distinguishing feature that would separate mathematical models from nonmathematical ones?

I would describe this feature by using such terms as autonomous, isolated, stable, self-contained, and—as a summary—formal. Autonomous and isolated—because mathematical models can be investigated "on their own" in isolation from the modeled objects. And one can do this for many years without any external information flow. Stable—because any modification of a mathematical model is qualified explicitly as defining a new model. No implicit modifications are allowed. Self-contained—because all properties of a mathematical model must be formulated explicitly. The term *formal model* can be used to summarize all these features.

For example, toy automobiles are autonomous, isolated, and stable models of "big" automobiles, but they are not self-contained because, as physical objects, toys possess a huge number of very complicated physical properties that a) are explained by complicated physical theories; b) do not play any role in modeling; c) are not separated explicitly from the properties really involved in modeling. Thus, to make our toy model self-contained, we should include (at least) quantum electrodynamics as part of it!

J. T.   I'm not sure I understand the analogy. First, do you mean that mathematical models should retain only essential features of the objects that they model? And second, by "self-contained," do you mean that these mathematical models should be complete within themselves in the same way, for example, that Euclidean geometry is complete? I mean that Euclidean geometry is what-ever can be deduced from the axioms, and if a result cannot be deduced from Euclid's axioms then it is not part of Euclidean geometry. In this sense, it is self-contained. Is this what you mean when you say mathematical models should be self-contained?

K. P.   As with most models, formal models may include ines-sential and even "wrong" properties. For example, many good models of the solar system represent planets not as massive bodies but "wrongly" as massive points. Thus, from the "goodness" point of view, mathematical models are as good or as bad as any other products of human intelligence.

Yes, indeed, the description of a self-contained model must include ALL assumptions that are allowed to derive new infor-mation (prove theorems) about the model. Thus, to make a self-contained model of a toy automobile, you must do one of two things: a) either separate explicitly which properties of the toy are included in the model (for example, if you are interested only in the shape of the vehicle, then declare this explicitly, scan the shape into your computer, and allow the use of analytical geometry to derive information); or b) include in the model all physical, chemi-cal, etc., theories necessary to draw conclusions about physical properties of the toy (for example, how would it behave under very high temperatures, high gamma radiation, etc.). Following the first way, you will obtain a simple mathematical model containing only a few (but almost only essential) properties of the vehicle. Following the second way, you would obtain a very complicated mathematical model, containing a huge number of inessential properties.

Now, the move from mathematical models to mathematics is as follows: For me, the task of mathematics is developing methods for creating and exploring mathematical models as defined above. As put by Morris Kline: "More than anything else mathematics is a method."

J. T.   So with respect to modeling sets, would you say that there is a sort of world of sets, and mathematicians develop mathematical models of this world? (This is the mathematician as an explorer of the mathematical landscape.) To use a specific example, would you call the theory of sets that arises from the axioms of Ernst Zermelo a mathematical model? That gets to the heart of the question. To quote what you said about physicists and chemists, is Zermelo's set theory a mathematical model for the mathematician's "part of the world"?

K. P.   In the philosophy of science, models and theories are treated as different categories. Theories are a means of model-building. For example, by using the theory of Newtonian mechanics with the Gravitation Law included, one can build models of various systems of "particles": planet systems, galaxies, etc.

From the axiom and theorem point of view, mathematical theories and models are very similar—any or both can be represented as a set of axioms and rules of inference allowing one to generate theorems.

Zermelo-Fraenkel set theory (ZFC) arose, indeed, as a model— the second attempt to describe the vision of "the world of sets" invented by Georg Cantor in the 1870s. The first attempt at axiomatic description failed. The simplest possible system of set axioms (in fact, a single axiom—the so-called unrestricted comprehension schema) leads quickly to contradictions (the famous Russell's paradox and some others). Is ZFC a "correct description" of Cantor's intuitive vision of sets? Or was Russell's paradox already "built" into Cantor's vision, and hence, ZFC represents a new and "better" version of the world of sets that is not identical to Cantor's world? Anyway, in ZFC, one can rebuild all of the common mathematics (all except some exotic highly theoretical results that need additional axioms, for example, the so-called large cardinal axioms).

Is ZFC a model of the mathematicians' "part of the world"? I would answer "no," it is not a model. It IS the mathematicians' part of the world; they do not know any better one.

J. T.   This seems inconsistent with what you said before. Do you mean that you think the ZFC model for sets is the best model that is currently available, or do you mean that it is an example of discovered final truth, a concept that you mentioned earlier?

K. P.   Your question, as well as my sudden turn "off the modeling" come close to the biggest controversy in the philosophy of mathematics. ZFC started, indeed, as an attempt to describe a vision of the world of sets. The unrestricted comprehension axiom schema led to paradoxes. So Zermelo introduced a restricted set of comprehension axioms that wouldn't allow for reproduction of the known paradoxes, but should be sufficient for the reproduction of theorems already proved about sets. Zermelo's idea was extremely successful. Even now, 100 years later, ZFC still dominates the market of set theories.

After this, should we still think of ZFC as a model of some more prominent structure that exists independently of the axioms of ZFC? Cantor's intuitive "world of sets with Russell's paradox inside" is not a good candidate for such a prominent structure. Thus, have mathematicians invented another world of sets, one that is better than Cantor's, and that is described correctly in the axioms of ZFC but exists independently of these axioms? Or has this "better world of sets" existed since the Big Bang, and mathematicians (starting with Cantor) have been trying to build a correct model of it?

This fantastic chain of questions can be answered in two ways. The minimalist way: cut the chain at the very beginning. ZFC, after being formulated, and after 100 years of continued success, does not need any more prominent structure behind it. The axioms of ZFC themselves ARE the best world of sets known to mathematicians. This point of view is called the "formalist philosophy of mathematics."

But there is also the maximalist way: Let us believe that, indeed, the "best" world of sets has existed since the Big Bang, and mathematicians are simply trying to build a correct model of it. This point of view is called the "Platonist philosophy of mathematics." (Plato introduced the "world of ideas" as something separate from the "world of things" 2,400 years ago.) At least until now, the so-called theory of large cardinals seems to support this point of view.

J. T. To make the discussion more concrete ... Bertrand Russell wrote a short article called "Definition of Number" in which he defines what is meant—or at least what he meant—by a natural number. In it, he describes the number 3 as something that all "trios" have in common. (When he says "trio," he means a set with three objects. Three particular people, three particular stones, and the set consisting of the words "paper, rock, scissors" are examples of trios.) Each such set is an "instance" of the number 3. When I read the article, I enjoyed it, and it made sense to me. But then I began to think about very large integers—integers, for example, that are much larger than the number of all the atoms in the universe. What would be an instance of this size number? And if there is no instance of such a large integer, in what sense does the integer exist?

K. P. Of course, the (now-called) natural numbers 1, 2, 3, . . ., billion, etc., arose from the human practice of counting. In mathematics, this human process of "number creation" ended with the axioms (for example, the so-called Peano axioms) describing the infinite natural number sequence as a whole. There is no problem with the existence of the axioms—one can write them down on paper. But what about the existence of the very very large numbers predicted by the axioms? According to the axioms, the number 10^{1000} can be obtained by adding 1 to 0 many times. But physicists know that the universe, as a computer, could not perform this "computation," even working continually since the Big Bang. Thus, the mathematical "world of numbers" is, in part, a kind of Disneyland—most really big numbers are of the Tom and Jerry kind.

J. T. What do you mean "the universe, as a computer"? And if large numbers are a sort of fiction, is it because they are too large to obtain by counting or because to the best of our knowledge no instance of such a number exists? For large enough numbers, of course, both properties must be true.

K. P. Of course, most probably, the universe is not a computer (at least not a usable one). But if you could imagine a computer as big as the universe, how many bits could it store, and how many operations could it have performed since the Big Bang? Physicists say no more than 10^{120} bits and no more than 10^{120} operations (Seth Lloyd)!

But if we represent numbers not as people of primitive times (as sequences of 1's), but as normal computers (i.e., in binary notation), then operating with numbers of size 10^{1000} is not a problem (just use 3,500 bits to represent a single number, and use the well-known simple algorithms to add and multiply such numbers).

J. T.   But no matter how large the largest numbers that can be stored within a computer, most numbers will be bigger still—

K. P.   Yes, for example, the "tower of four tens"—$10^{\wedge}(10^{\wedge}(10^{\wedge}10))$, where $\wedge$ stands for the power operation—never will be represented either in the binary or in the decimal notation. We can operate with such "numbers" only in a very limited sense. Aren't we, in fact, operating with number expressions rather than with numbers?

J. T.   Another way of thinking about the natural numbers is that they are "closed under addition." Most people accept the reality of small natural numbers, and they accept the requirement that it is always possible to add 1 to a natural number to obtain the natural number that is "the next one over," but then the larger natural numbers must exist, because they are logical consequences of this closure requirement. We only need to begin with 1 and then we just add 1 until we have generated large natural numbers. But this, it seems to me, is closer to Aristotle's idea of the infinite. In his book *Physics* he wrote about the infinite in geometry. He said, "In point of fact they *[mathematicians]* do not need the infinite and do not use it. They postulate only that the finite straight line may be produced as far as they wish." What do you think?

K. P.   Most mathematicians do not agree with Aristotle, and they use the Axiom of Infinity to obtain big and bigger actually infinite sets. But, of course, Aristotle was a brilliant thinker of his time, and his idea that, in fact, mathematicians do not need the actual infinite (only the potential one) is not completely wrong. Moreover, today, we know (as you say, "to the best of our knowledge") that the finite straight line CANNOT be produced as far as we wish—in the universe, because of gravity, there are no very long straight lines at all. Potentially infinite straight lines are idealizations, but they appear to be very good for building useful mathematical models and—to some extent—the same is true of actually infinite sets.

J. T.   How have your ideas about the reality of mathematics affected your own mathematical research?

K. P.   Unfortunately, I left mathematics for computer science at the age of 35 (now I'm 60). My recent 25-year experience includes the theory and practice of computer programming, database design, graphical tool development for business system modeling, etc. If I would be allowed to carry out mathematical research, I would try to build a new arithmetic that would use arithmetical expressions and not numbers as the fundamental notion. But it seems I won't, so I would invite younger people to try this idea that was inspired by my lifelong philosophical development.

There is another philosophical idea that I would be happy to develop mathematically. Reading Henri Poincaré, I realized that arithmetic "should be" inconsistent, i.e., there should be a way to derive a contradiction from the axioms of arithmetic. The idea is as follows. In trying to axiomatize the notion of natural numbers, we are building a vicious circle: The notion of proof from the axioms includes the so-called mathematical induction, but this induction also represents the main feature of the natural number system that we are trying to axiomatize.

The most serious partial results in this direction were obtained by Edward Nelson. But, if we try searching the Web for possible contradictions in mathematics, then we can find a serious announcement by Nikolai Belyakin: If we add to Zermelo-Fraenkel set theory the second weakest large cardinal axiom, then we obtain a contradiction. However, the full proof of this result is not yet published.

Of course, an inconsistency proof of arithmetic will not put an end to the banking business. Nor will this mean that Intel processors are built on a "wrong theory." No harm will be done to applications of mathematics because it is only the "Tom and Jerry part" of arithmetic that "should be" inconsistent!

J. T.   Mathematics seems to be a sort of cross-cultural language. Of course, there are people who find math inaccessible, but mathematicians from around the world usually seem to agree on when a theorem has been proved. This is remarkable to me because mathematicians often share no common spoken language and have very different cultural backgrounds. Depending on their backgrounds, they may approach mathematics in different ways, but they still agree on the main points—at least that is how it seems to me. Do you agree, and if so, do you think that this reveals more about how the human brain works than about anything that mathematics purports to describe?

K. P.   As I have been trying to promote for many years, the task of mathematics is developing methods of creating and exploring mathematical models (in the broadest possible sense). Are the general features of the "world of all the possible mathematical models" determined by the features of how the human brain works or by the features of how the physical world is or both? Could an alien civilization design its world of mathematical models in a radically different way from our way? For example, would they use the same kind of natural numbers that we are using? I guess that the answer should be "yes." But could we try proving this as a mathematical theorem? Would it be a theorem of our mathematics or theirs?

J. T.   Thank you for sharing your considerable insight into the nature of mathematics. I've enjoyed our conversation.

CHRONOLOGY

ca. 3000 B.C.E.
Hieroglyphic numerals are in use in Egypt.

ca. 2500 B.C.E.
Construction of the Great Pyramid of Khufu takes place.

ca. 2400 B.C.E.
An almost complete system of positional notation is in use in Mesopotamia.

ca. 1650 B.C.E.
The Egyptian scribe Ahmes copies what is now known as the Ahmes (or Rhind) papyrus from an earlier version of the same document.

ca. 585 B.C.E.
Thales of Miletus carries out his research into geometry, marking the beginning of mathematics as a deductive science.

ca. 540 B.C.E.
Pythagoras of Samos establishes the Pythagorean school of philosophy.

ca. 500 B.C.E.
Rod numerals are in use in China.

ca. 420 B.C.E.
Zeno of Elea proposes his philosophical paradoxes.

ca. 399 B.C.E.
Socrates dies.

ca. 360 B.C.E.
Eudoxus, author of the method of exhaustion, carries out his research into mathematics.

ca. 350 B.C.E.

The Greek mathematician Menaechmus writes an important work on conic sections.

ca. 347 B.C.E.

Plato dies.

332 B.C.E.

Alexandria, Egypt, center of Greek mathematics, is established.

ca. 300 B.C.E.

Euclid of Alexandria writes *Elements*, one of the most influential mathematics books of all time.

ca. 260 B.C.E.

Aristarchus of Samos discovers a method for computing the ratio of the Earth-Moon distance to the Earth-Sun distance.

ca. 230 B.C.E.

Eratosthenes of Cyrene computes the circumference of Earth.

Apollonius of Perga writes *Conics*.

Archimedes of Syracuse writes *The Method*, *Equilibrium of Planes*, and other works.

206 B.C.E.

The Han dynasty is established; Chinese mathematics flourishes.

ca. C.E. 150

Ptolemy of Alexandria writes *Almagest*, the most influential astronomy text of antiquity.

ca. C.E. 250

Diophantus of Alexandria writes *Arithmetica*, an important step forward for algebra.

ca. 320

Pappus of Alexandria writes his *Collection*, one of the last influential Greek mathematical treatises.

415

The death of the Alexandrian philosopher and mathematician Hypatia marks the end of the Greek mathematical tradition.

ca. 476

The astronomer and mathematician Aryabhata is born; Indian mathematics flourishes.

ca. 630

The Hindu mathematician and astronomer Brahmagupta writes *Brahma Sphuta Siddhānta*, which contains a description of place-value notation.

ca. 775

Scholars in Baghdad begin to translate Hindu and Greek works into Arabic.

ca. 830

Mohammed ibn-Mūsā al-Khwārizmī writes *Hisāb al-jabr wa'l muqābala*, a new approach to algebra.

833

Al-Ma'mūn, founder of the House of Wisdom in Baghdad, Iraq, dies.

ca. 840

The Jainist mathematician Mahavira writes *Ganita Sara Samgraha*, an important mathematical textbook.

1086

An intensive survey of the wealth of England is carried out and summarized in the tables and lists of the *Domesday Book*.

1123

Omar Khayyám, the author of *Al-jabr w'al muqābala* and the *Rubái-yát*, the last great classical Islamic mathematician, dies.

ca. 1144

Bhaskara II writes the *Lilavati* and the *Vija-Ganita*, two of the last great works in the classical Indian mathematical tradition.

ca. 1202

Leonardo of Pisa (Fibonacci), author of *Liber Abaci*, arrives in Europe.

1360

Nicholas Oresme, a French mathematician and Roman Catholic bishop, represents distance as the area beneath a velocity line.

1471

The German artist Albrecht Dürer is born.

1482

Leonardo da Vinci begins to keep his diaries.

ca. 1541

Niccolò Fontana, an Italian mathematician, also known as Tartaglia, discovers a general method for factoring third-degree algebraic equations.

1543

Copernicus publishes *De revolutionibus*, marking the start of the Copernican revolution.

1545

Girolamo Cardano, an Italian mathematician and physician, publishes *Ars magna*, marking the beginning of modern algebra. Later he publishes *Liber de ludo aleae*, the first book on probability.

1579

François Viète, a French mathematician, publishes *Canon mathematicus*, marking the beginning of modern algebraic notation.

1585

The Dutch mathematician and engineer Simon Stevin publishes "La disme."

1609

Johannes Kepler, author of Kepler's laws of planetary motion, publishes *Astronomia Nova*.

Galileo Galilei begins his astronomical observations.

1621

The English mathematician and astronomer Thomas Harriot dies. His only work, *Artis analyticae praxis*, is published in 1631.

ca. 1630

The French lawyer and mathematician Pierre de Fermat begins a lifetime of mathematical research. He is the first person to claim to have proved "Fermat's last theorem."

1636

Gérard (or Girard) Desargues, a French mathematician and engineer, publishes *Traité de la section perspective*, which marks the beginning of projective geometry.

1637

René Descartes, a French philosopher and mathematician, publishes *Discours de la méthode*, permanently changing both algebra and geometry.

1638

Galileo Galilei publishes *Dialogues Concerning Two New Sciences* while under arrest.

1640

Blaise Pascal, a French philosopher, scientist, and mathematician, publishes *Essai sur les coniques*, an extension of the work of Desargues.

1642

Blaise Pascal manufactures an early mechanical calculator, the Pascaline.

1654

Pierre de Fermat and Blaise Pascal exchange a series of letters about probability, thereby inspiring many mathematicians to study the subject.

1655

John Wallis, an English mathematician and clergyman, publishes *Arithmetica infinitorum*, an important work that presages calculus.

1657

Christiaan Huygens, a Dutch mathematician, astronomer, and physicist, publishes *De ratiociniis in aleae ludo*, a highly influential text in probability theory.

1662

John Graunt, an English businessman and a pioneer in statistics, publishes his research on the London Bills of Mortality.

1673

Gottfried Leibniz, a German philosopher and mathematician, constructs a mechanical calculator that can perform addition, subtraction, multiplication, division, and extraction of roots.

1683

Seki Kōwa, Japanese mathematician, discovers the theory of determinants.

1684

Gottfried Leibniz publishes the first paper on calculus, *Nova methodus pro maximis et minimis*.

1687

Isaac Newton, a British mathematician and physicist, publishes *Philosophiae naturalis principia mathematica*, beginning a new era in science.

1693

Edmund Halley, a British mathematician and astronomer, undertakes a statistical study of the mortality rate in Breslau, Germany.

1698

Thomas Savery, an English engineer and inventor, patents the first steam engine.

1705

Jacob Bernoulli, a Swiss mathematician, dies. His major work on probability, *Ars conjectandi*, is published in 1713.

1712

The first Newcomen steam engine is installed.

1718

Abraham de Moivre, a French mathematician, publishes *The Doctrine of Chances*, the most advanced text of the time on the theory of probability.

1743

The Anglo-Irish Anglican bishop and philosopher George Berkeley publishes *The Analyst*, an attack on the new mathematics pioneered by Isaac Newton and Gottfried Leibniz.

The French mathematician and philosopher Jean Le Rond d'Alembert begins work on the *Encyclopédie*, one of the great works of the Enlightenment.

1748

Leonhard Euler, a Swiss mathematician, publishes his *Introductio*.

1749

The French mathematician and scientist George-Louis Leclerc, count de Buffon publishes the first volume of *Histoire naturelle*.

1750

Gabriel Cramer, a Swiss mathematician, publishes "Cramer's rule," a procedure for solving systems of linear equations.

1760

Daniel Bernoulli, a Swiss mathematician and scientist, publishes his probabilistic analysis of the risks and benefits of variolation against smallpox.

1761

Thomas Bayes, an English theologian and mathematician, dies. His "Essay Towards Solving a Problem in the Doctrine of Chances" is published two years later.

The English scientist Joseph Black proposes the idea of latent heat.

1769

James Watt obtains his first steam engine patent.

1781

William Herschel, a German-born British musician and astronomer, discovers Uranus.

1789

Unrest in France culminates in the French Revolution.

1793

The Reign of Terror, a period of brutal, state-sanctioned repression, begins in France.

1794

The French mathematician Adrien-Marie Legendre (or Le Gendre) publishes his *Éléments de géométrie*, a text that influences mathematics education for decades.

Antoine-Laurent Lavoisier, a French scientist and discoverer of the law of conservation of mass, is executed by the French government.

1798

Benjamin Thompson (Count Rumford), a British physicist, proposes the equivalence of heat and work.

1799

Napoléon seizes control of the French government.

Caspar Wessel, a Norwegian mathematician and surveyor, publishes the first geometric representation of the complex numbers.

1801

Carl Friedrich Gauss, a German mathematician, publishes *Disquisitiones arithmeticae*.

1805

Adrien-Marie Legendre, a French mathematician, publishes *Nouvelles méthodes pour la détermination des orbites des comètes*, which contains the first description of the method of least squares.

1806

Jean-Robert Argand, a French bookkeeper, accountant, and mathematician, develops the Argand diagram to represent complex numbers.

1812

Pierre-Simon Laplace, a French mathematician, publishes *Theorie analytique des probabilités*, the most influential 19th-century work on the theory of probability.

1815

Napoléon suffers final defeat at the battle of Waterloo.

Jean-Victor Poncelet, a French mathematician and the "father of projective geometry," publishes *Traité des propriétés projectives des figures.*

1824

The French engineer Sadi Carnot publishes *Réflexions sur la puissance motrice du feu,* wherein he describes the Carnot engine.

Niels Henrik Abel, a Norwegian mathematician, publishes his proof of the impossibility of algebraically solving a general fifth-degree equation.

1826

Nikolay Ivanovich Lobachevsky, a Russian mathematician and "the Copernicus of geometry," announces his theory of non-Euclidean geometry.

1828

Robert Brown, a Scottish botanist, publishes the first description of Brownian motion in "A Brief Account of Microscopical Observations."

1830

Charles Babbage, a British mathematician and inventor, begins work on his analytical engine, the first attempt at a modern computer.

1832

János Bolyai, a Hungarian mathematician, publishes *Absolute Science of Space.*

The French mathematician Évariste Galois is killed in a duel.

1843

James Prescott Joule publishes his measurement of the mechanical equivalent of heat.

1846

The planet Neptune is discovered by the French mathematician Urbain-Jean-Joseph Le Verrier from a mathematical analysis of the orbit of Uranus.

1847

Georg Christian von Staudt publishes *Geometrie der Lage*, which shows that projective geometry can be expressed without any concept of length.

1848

Bernhard Bolzano, a Czech mathematician and theologian, dies. His study of infinite sets, *Paradoxien des Unendlichen*, is first published in 1851.

1850

Rudolph Clausius, a German mathematician and physicist, publishes his first paper on the theory of heat.

1851

William Thomson (Lord Kelvin), a British scientist, publishes "On the Dynamical Theory of Heat."

1854

George Boole, a British mathematician, publishes *Laws of Thought*. The mathematics contained therein makes possible the later design of computer logic circuits.

The German mathematician Bernhard Riemann gives the historic lecture "On the Hypotheses That Form the Foundations of Geometry." The ideas therein play an integral part in the theory of relativity.

1855

John Snow, a British physician, publishes "On the Mode of Communication of Cholera," the first successful epidemiological study of a disease.

1859

James Clerk Maxwell, a British physicist, proposes a probabilistic model for the distribution of molecular velocities in a gas.

Charles Darwin, a British biologist, publishes *On the Origin of Species by Means of Natural Selection*.

1861

Karl Weierstrass creates a continuous nowhere differentiable function.

1866

The Austrian biologist and monk Gregor Mendel publishes his ideas on the theory of heredity in "Versuche über Pflanzenhybriden."

1872

The German mathematician Felix Klein announces his Erlanger Programm, an attempt to categorize all geometries with the use of group theory.

Lord Kelvin (William Thomson) develops an early analog computer to predict tides.

Richard Dedekind, a German mathematician, rigorously establishes the connection between real numbers and the real number line.

1874

Georg Cantor, a German mathematician, publishes "Über eine Eigenschaft des Inbegriffes aller reelen algebraischen Zahlen," a pioneering paper that shows that all infinite sets are not the same size.

1890

The Hollerith tabulator, an important innovation in calculating machines, is installed at the United States Census for use in the 1890 census.

Giuseppe Peano publishes his example of a space-filling curve.

1894

Oliver Heaviside describes his operational calculus in his text *Electromagnetic Theory*.

1895

Henri Poincaré publishes *Analysis situs*, a landmark paper in the history of topology, in which he introduces a number of ideas that would occupy the attention of mathematicians for generations.

1898

Émile Borel begins to develop a theory of measure of abstract sets that takes into account the topology of the sets on which the measure is defined.

1899

The German mathematician David Hilbert publishes the definitive axiomatic treatment of Euclidean geometry.

1900

David Hilbert announces his list of mathematics problems for the 20th century.

The Russian mathematician Andrey Andreyevich Markov begins his research into the theory of probability.

1901

Henri-Léon Lebesgue, a French mathematician, develops his theory of integration.

1905

Ernst Zermelo, a German mathematician, undertakes the task of axiomatizing set theory.

Albert Einstein, a German-born American physicist, begins to publish his discoveries in physics.

1906

Marian Smoluchowski, a Polish scientist, publishes his insights into Brownian motion.

1908

The Hardy-Weinberg law, containing ideas fundamental to population genetics, is published.

1910

Bertrand Russell, a British logician and philosopher, and Alfred North Whitehead, a British mathematician and philosopher, publish *Principia mathematica*, an important work on the foundations of mathematics.

1913

Luitzen E. J. Brouwer publishes his recursive definition of the concept of dimension.

1914

Felix Hausdorff publishes *Grundzüge der Mengenlehre.*

1915

Wacław Sierpiński publishes his description of the now-famous curve called the Sierpiński gasket.

1917

Vladimir Ilyich Lenin leads a revolution that results in the founding of the Union of Soviet Socialist Republics.

1918

World War I ends.

The German mathematician Emmy Noether presents her ideas on the roles of symmetries in physics.

1920

Zygmunt Janiszewski, founder of the Polish school of topology, dies.

1923

Stefan Banach begins to develop the theory of Banach spaces.

Karl Menger publishes his first paper on dimension theory.

1924

Pavel Samuilovich Urysohn dies in a swimming accident at the age of 25 after making several important contributions to topology.

1928

Maurice Frechet publishes his *Les espaces abstraits et leurs théorie considérée comme introduction à l'analyse générale*, which places topological concepts at the foundation of the field of analysis.

1929

Andrey Nikolayevich Kolmogorov, a Russian mathematician, publishes *General Theory of Measure and Probability Theory*, establishing the theory of probability on a firm axiomatic basis for the first time.

1930

Ronald Aylmer Fisher, a British geneticist and statistician, publishes *Genetical Theory of Natural Selection,* an important early attempt to express the theory of natural selection in mathematical language.

1931

Kurt Gödel, an Austrian-born American mathematician, publishes his incompleteness proof.

The Differential Analyzer, an important development in analog computers, is developed at Massachusetts Institute of Technology.

1933

Karl Pearson, a British innovator in statistics, retires from University College, London.

Kazimierz Kuratowski publishes the first volume of *Topologie,* which extends the boundaries of set theoretic topology (still an important text).

1935

George Horace Gallup, a U.S. statistician, founds the American Institute of Public Opinion.

1937

The British mathematician Alan Turing publishes his insights on the limits of computability.

Topologist and teacher Robert Lee Moore begins serving as president of the American Mathematical Society.

1939

World War II begins.

William Edwards Deming joins the United States Census Bureau.

The Nicholas Bourbaki group publishes the first volume of its *Éléments de mathématique.*

Sergei Sobolev elected to the USSR Academy of Sciences after publishing a long series of papers describing a generalization of the concept of function and a generalization of the concept of derivative. His work forms the foundation for a new branch of analysis.

1941

Witold Hurewicz and Henry Wallman publish their classic text *Dimension Theory*.

1945

Samuel Eilenberg and Saunders MacLane found the discipline of category theory.

1946

The Electronic Numerical Integrator and Calculator (ENIAC) computer begins operation at the University of Pennsylvania.

1948

While working at Bell Telephone Labs in the United States, Claude Shannon publishes "A Mathematical Theory of Communication," marking the beginning of the Information Age.

1951

The Universal Automatic Computer (UNIVAC I) is installed at U.S. Bureau of the Census.

1954

FORmula TRANslator (FORTRAN), one of the first high-level computer languages, is introduced.

1956

The American Walter Shewhart, an innovator in the field of quality control, retires from Bell Telephone Laboratories.

1957

Olga Oleinik publishes "Discontinuous Solutions to Nonlinear Differential Equations," a milestone in mathematical physics.

1965

Andrey Nikolayevich Kolmogorov establishes the branch of mathematics now known as Kolmogorov complexity.

1972

Amid much fanfare, the French mathematician and philosopher René Thom establishes a new field of mathematics called catastrophe theory.

1973

The C computer language, developed at Bell Laboratories, is essentially completed.

1975

The French geophysicist Jean Morlet helps develop a new kind of analysis based on what he calls "wavelets."

1980

Kiiti Morita, the founder of the Japanese school of topology, publishes a paper that further extends the concept of dimension to general topological spaces.

1982

Benoît Mandelbrot publishes his highly influential *The Fractal Geometry of Nature*.

1989

The Belgian mathematician Ingrid Daubechies develops what has become the mathematical foundation for today's wavelet research.

1995

The British mathematician Andrew Wiles publishes the first proof of Fermat's last theorem.

JAVA computer language is introduced commercially by Sun Microsystems.

1997

René Thom declares the mathematical field of catastrophe theory "dead."

2002

Experimental Mathematics celebrates its 10th anniversary. It is a refereed journal dedicated to the experimental aspects of mathematical research.

Manindra Agrawal, Neeraj Kayal, and Nitin Saxena create a brief, elegant algorithm to test whether a number is prime, thereby solving an important centuries-old problem.

2003

Grigory Perelman produces the first complete proof of the Poincaré conjecture, a statement about some of the most fundamental properties of three-dimensional shapes.

2007

The international financial system, heavily dependent on so-called sophisticated mathematical models, finds itself on the edge of collapse, calling into question the value of the mathematical models.

2008

Henri Cartan, one of the founding members of the Nicholas Bourbaki group, dies at the age of 104.

GLOSSARY

algebraic equation an equation of the form $a_n x^n + a_{n-1}x^{n-1} + \ldots + a_1 x + a_0 = 0$, where n can represent any natural number, x represents the variable raised to the power indicated, and a_j, which always denotes a rational number, is the coefficient by which x^j is multiplied

axiom a statement accepted as true that serves as a basis for deductive reasoning

axiomatize to develop a set of axioms for a mathematical system

cardinal number the property that a set has in common with all sets equivalent to it. The cardinal number of a finite set is simply the number of elements in the set. For an infinite set the cardinal number is often used to identify a reference set with which the given set can be placed in one-to-one correspondence

coefficient a number or symbol used to multiply a variable

complete (set of axioms) a set of axioms defining a mathematical system with the property that every statement in the system can be proved either true or false

complex number any number of the form $a + bi$ where a and b are real numbers and i has the property that $i^2 = -1$

consistent (set of axioms) a set of axioms from which it is not possible to deduce a theorem that is both true and false

convergent sequence a sequence of real numbers is said to converge provided there exists a real number L with the property that every circle containing L in its interior also contains all but finitely many terms of the sequence

countable set an infinite set for which there exists a 1-1 correspondence between its members and the set of natural numbers

deduction   a conclusion obtained by logically reasoning from general principles to particular statements

degree of an equation   for an algebraic equation of one variable the largest exponent appearing in the equation

extraction of roots   the procedure whereby given a number, here represented by the letter p, another number written $\sqrt[n]{p}$, the nth root of p, is computed. A number $\sqrt[n]{p}$ has the property that $(\sqrt[n]{p})^n = p$

fixed-point arithmetic   any computational procedure in which one computes with numbers composed of $t + s$ digits—where t digits are used to represent the integer part of each number and s digits are used to represent the fractional part

floating-point arithmetic   any arithmetic procedure performed with numbers that are in normalized floating-point form. In base 10 that means any number, p, represented in the form $p = m \times 10^n$ where n is an integer and m is a number consisting of a fixed number of digits and $1 \leq m < 10$. Similar representations exist for other bases

halting problem   a problem, conceived by Alan Turing, that cannot be solved by any type of automatic computation. The problem itself involves predicting whether or not a computer will reach the end of a particular type of computation (and halt) or simply continue to compute forever

incomplete (set of axioms)   a set of axioms describing a mathematical system in which there exists a statement within the system that cannot be proved either true or false as a logical consequence of the axioms

infinite set   a set that can be placed in one-to-one correspondence with a proper subset of itself

integer   any number belonging to the set $\{0, -1, 1, -2, 2, -3, 3, \ldots\}$

irrational number   any real number that cannot be written as a/b where a and b are integers and b is not 0

one-to-one correspondence   the pairing of elements between two sets, A and B, so that each element of A is paired with a unique element of B and to each element of B is paired a unique element of A

perfect square   an integer p with the property that there exists another integer q such that $q^2 = p$

positional numeration   a system of notation in which each number is represented by a string of digits and the value of a digit depends on its position in the string

power set   given a set S, the power set of S is defined as the set whose elements are the subsets of S

proper subset   given two sets A and B, B is a proper subset of A if every element of B belongs to A but not every element of A belongs to B

quadratic equation   an algebraic equation in which the highest exponent appearing is of degree 2

quadratic formula   a mathematical formula for computing the roots of any quadratic equation by using the coefficients that appear in the equation

rational number   any number of the form a/b where a and b are integers and b is not 0

real number   any number that corresponds to a point on the real number line

root (of an equation)   any number that satisfies an algebraic equation

Russell paradox   a logical paradox that arises when a set is defined so that it is an element of itself. The Russell paradox shows that the "set of all sets" does not exist

sequence (of real numbers)   a set of real numbers that is placed in correspondence with the set of natural numbers

set   a collection of numbers, points, or other elements

sexagesimal   base 60

subset   given two sets, A and B, B is said to be a subset of A if every element of B also belongs to A

theorem   a statement that is deduced from the axioms that define a given mathematical system

transcendental number   a number that is not the root of any algebraic equation

transfinite number   the cardinal number of an infinite set

Turing machine   a theoretical machine used to pose and solve certain problems in computer science

uncomputable number   a number is said to be uncomputable whenever the shortest computer program needed to express the first n binary digits of the number is at least n binary digits long, and this statement holds for every natural number n

uncountable set   an infinite set for which there exists no 1-1 correspondence between the natural numbers and the elements of the set in question

unit fraction   a fraction of the form $1/a$ where a is any integer except 0

vigesimal   base 20

MODERN WORKS

Aczel, Amir D. *The Mystery of the Aleph: Mathematics, the Kaballah, and the Human Mind.* New York: Four Walls Eight Windows, 2000. This is the story of how mathematicians developed the concept of infinity. It also covers the religious motivations of Georg Cantor, the founder of modern set theory.

Adler, Irving. *Thinking Machines, a Layman's Introduction to Logic, Boolean Algebra, and Computers.* New York: John Day, 1961. This old book is still the best nontechnical introduction to computer arithmetic. It begins with fingers and ends with Boolean logic circuits.

Boyer, Carl B., and Uta C. Merzbach. *A History of Mathematics.* New York: John Wiley & Sons, 1991. Boyer was one of the preeminent mathematics historians of the 20th century. This work contains much interesting biographical information. The mathematical information assumes a fairly strong background.

Bruno, Leonard C. *Math and Mathematicians: The History of Mathematics Discoveries around the World,* 2 vols. Michigan: U·X·L, 1999. Despite its name there is little mathematics in this two-volume set. What you find is a very large number of brief biographies of many individuals who were important in the history of mathematics.

Bunt, Lucas Nicolaas Hendrik, Phillip S. Jones, Jack D. Bedient. *The Historical Roots of Elementary Mathematics.* Englewood Cliffs, N.J.: Prentice Hall, 1976. A highly detailed examination—complete with numerous exercises—of how ancient cultures added, subtracted, divided, multiplied, and reasoned.

Courant, Richard, and Herbert Robbins. *What Is Mathematics? An Elementary Approach to Ideas and Mathematics.* New York: Oxford University Press, 1941. A classic and exhaustive answer to the question posed in the title. Courant was an influential 20th-century mathematician.

Danzig, Tobias. *Number, the Language of Science*. New York: Macmillan, 1954. First published in 1930, this book is painfully elitist; the author's many prejudices are on display in every chapter. Yet it is one of the best nontechnical histories of the concept of number ever written. Apparently it was also Albert Einstein's favorite book on the history of mathematics.

Davis, Phillip J. *The Lore of Large Numbers*. New York: Random House, 1961. An excellent overview of numbers, ways they are written, and ways they are used in science.

Dewdney, Alexander K. *200% of Nothing: An Eye-Opening Tour through the Twists and Turns of Math Abuse and Innumeracy*. New York: John Wiley & Sons, 1993. A critical look at ways mathematical reasoning has been abused to distort truth.

Eastaway, Robert, and Jeremy Wyndham. *Why Do Buses Come in Threes? The Hidden Mathematics of Everyday Life*. New York: John Wiley & Sons, 1998. Nineteen lighthearted essays on the mathematics underlying everything from luck to scheduling problems.

Eves, Howard. *An Introduction to the History of Mathematics*. New York: Holt, Rinehart & Winston, 1953. This well-written history of mathematics places special emphasis on early mathematics. It is unusual because the history is accompanied by numerous mathematical problems. (The solutions are in the back of the book.)

Gardner, Martin. *The Colossal Book of Mathematics*. New York: Norton, 2001. Martin Gardner had a gift for seeing things mathematically. This "colossal" book contains sections on geometry, algebra, probability, logic, and more.

———. *Logic Machines and Diagrams*. Chicago: University of Chicago Press, 1982. An excellent book on logic and its uses in computers.

Guillen, Michael. *Bridges to Infinity: The Human Side of Mathematics*. Los Angeles: Jeremy P. Tarcher, 1983. This book consists of an engaging nontechnical set of essays on mathematical topics, including non-Euclidean geometry, transfinite numbers, and catastrophe theory.

Hoffman, Paul. *Archimedes' Revenge: The Joys and Perils of Mathematics*. New York: Ballantine, 1989. A relaxed, sometimes silly look at an interesting and diverse set of math problems ranging from prime numbers and cryptography to Turing machines and the mathematics of democratic processes.

Joseph, George G. *The Crest of the Peacock: The Non-European Roots of Mathematics*. Princeton, N.J.: Princeton University Press, 1991. One of the best of a new crop of books devoted to this important topic.

Kaplan, Robert. *The Nothing That Is: A Natural History of Zero*. New York: Oxford University Press, 1999. Zero was one of the great conceptual breakthroughs in the history of mathematics. This is a lively history devoted to that one concept.

Kasner, Edward, and James R. Newman. "Paradox Lost and Paradox Regained." In *The World of Mathematics*. Vol. 3. New York: Dover Publications, 1956. A discussion of many common mathematical paradoxes and their logical resolutions.

Kline, Morris. *Mathematics for the Nonmathematician*. New York: Dover Publications, 1985. An articulate, not very technical overview of many important mathematical ideas.

———. *Mathematics in Western Culture*. New York: Oxford University Press, 1953. An excellent overview of the development of Western mathematics in its cultural context, this book is aimed at an audience with a firm grasp of high school–level mathematics.

Maor, Eli. *e: The Story of a Number*. New York: Oxford University Press, 1999. The transcendental number called *e* plays a peculiar and very important role in calculus and the mathematics that grew out of calculus. This is an entire book dedicated to telling the story of *e*.

McLeish, John. *Number*. New York: Fawcett Columbine, 1992. A history of the concept of number from Mesopotamian to modern times.

Niven, Ivan Morton. *Numbers, Rational and Irrational.* New York: Random House, 1961. Aimed at older high school students, this is a challenging mathematical presentation of our number system.

Pappas, Theoni. *The Joy of Mathematics.* San Carlos, Calif.: World Wide/Tetra, 1986. Aimed at a younger audience, this work searches for interesting applications of mathematics in the world around us.

Pierce, John R. *An Introduction to Information Theory: Symbols, Signals and Noise.* New York: Dover Publications, 1961. Despite the sound of the title, this is not a textbook. Pierce, formerly of Bell Laboratories, describes how (among other topics) numbers and text are digitally encoded for transmission and storage—a lucid introduction to a very important topic.

Reid, Constance. *From Zero to Infinity: What Makes Numbers Interesting.* New York: Thomas Y. Crowell, 1960. A well-written overview of numbers and the algebra that stimulated their development.

Seife, Charles. *Zero: The Biography of a Dangerous Idea.* New York: Viking Press, 2000. A book dedicated to one of the most productive ideas in the history of mathematics.

Smith, David E., and Yoshio Mikami. *A History of Japanese Mathematics.* Chicago: Open Court, 1914. Copies of this book are still around, and it is frequently quoted. The first half is an informative nontechnical survey. The second half is written more for the expert.

Stewart, Ian. *From Here to Infinity.* New York: Oxford University Press, 1996. A well-written, very readable overview of several important contemporary ideas in geometry, algebra, computability, chaos, and mathematics in nature.

Swetz, Frank J., editor. *From Five Fingers to Infinity: A Journey through the History of Mathematics.* Chicago: Open Court, 1994. This is a fascinating, though not especially focused, look at the history of mathematics. Highly recommended.

Tabak, John. *Algebra. History of Mathematics.* Rev. ed. New York: Facts On File, 2004. More information about how the concept of number and ideas about the nature of algebra evolved together.

Thomas, David A. *Math Projects for Young Scientists.* New York: Franklin Watts, 1988. This project-oriented text gives an introduction to several historically important geometry problems.

White, Leslie. "The Locus of Mathematical Reality: An Anthropological Footnote." In *The World of Mathematics.* Vol. 4, edited by James R. Newman. New York: Dover Publications, 1956. Leslie White seeks to show that "mathematics is nothing more than a particular kind of primate behavior." This article is a well-written, accessible, and interesting discussion of the nature of mathematics.

Zippin, Leo. *The Uses of Infinity.* New York: Random House, 1962. Contains lots of equations—perhaps too many for the uninitiated—but none of the equations is very difficult. The book is worth the effort needed to read it.

ORIGINAL SOURCES

Reading the discoverer's own description can sometimes deepen our appreciation of an important mathematical discovery. Often this is not possible, because the description is too technical. Fortunately there are exceptions. Sometimes the discovery is accessible because the idea does not require a lot of technical background to be appreciated. Sometimes the discoverer writes a nontechnical account of the technical idea that she or he has discovered. Here are some classic papers:

Ahmes. *The Rhind Mathematical Papyrus: Free Translation, Commentary, and Selected Photographs, Transcription, Literal Translations.* Translated by Arnold B. Chace. Reston, Va.: National Council of Teachers of Mathematics, 1979. This is a translation of the longest and best of extant Egyptian mathematical texts, the Rhind papyrus (also known as the Ahmes papyrus). It provides insight into the types of problems and methods of solution known to one of humanity's oldest cultures.

Alexandrov, A. D., A. N. Kolmogorov, and M. A. Lavrent'ev, eds. *Mathematics, Its Content, Methods, and Meaning.* Translated by S. H. Gould and T. Bartha. Mineola, N.Y.: Dover Publications, 1999. First published in English in the early 1960s in three volumes, this overview of contemporary mathematics, pure and applied, was written by some of the best mathematicians of their era.

Chaitin, Gregory. *The Limits of Mathematics: A Course on Information Theory and Limits of Formal Reasoning.* New York: Springer, 1998. Written for a lay audience, this is Chaitin's account of what he believes is a revolution in the meaning and future of mathematics.

Galilei, Galileo. *Dialogues Concerning Two New Sciences.* Translated by Henry Crew and Alphonso de Salvio. New York: Dover Publications, 1954. Galileo's discussion of infinite sets begins on page 31 of this edition. It is fascinating to read his grappling with this difficult subject, and as he struggles with the idea of infinity he sees further into the subject than any of his predecessors.

Hahn, Hans. Infinity. In *The World of Mathematics.* Vol. 3, edited by James Newman. New York: Dover Publications, 1956. A thoughtful exposition of some of the fundamental ideas involved in the study of infinite sets by one of the great mathematicians of the first half of the 20th century.

Hardy, Godfrey H. *A Mathematician's Apology.* Cambridge, England: Cambridge University Press, 1940. Hardy was an excellent mathematician and a good writer. In this oft-quoted and very brief book Hardy seeks to explain and sometimes justify his life as a mathematician.

Russell, Bertrand. Definition of Number. In *The World of Mathematics.* Vol. 1, edited by James R. Newman. New York: Dover Publications, 1956. Russell was one of the great thinkers of the 20th century and a pioneer in mathematics. Here he describes for a general audience contemporary insights into the nature of number.

———. "Mathematics and the Metaphysicians." In *The World of Mathematics.* Vol. 3, edited by James Newman. New York: Dover Publications, 1956. An introduction to the philosophical ideas on which mathematics is founded written by a major contributor to this field.

Stevin, Simon. *Principal Works*, edited by Ernes Crane, E. J. Dijk-sterhuis, A. D. Fokker, R. J. Forbes, A. Pannekock, A. Romein-Verschoor, W. H. Schukking, D. J. Struik. Translated by C. Dikshoorn. 5 vols. Amsterdam: C. V. Swets and Zeitlinger, 1955–1966. Every history of mathematics devotes space to Simon Stevin, but unlike those of Galileo and Kepler, Stevin's writings are difficult to find. This very readable translation is available in some larger libraries.

Turing, Alan M. "Can a Machine Think?" In *The World of Mathematics*. Vol. 4, edited by James R. Newman. New York: Dover, 1956. In many ways modern theoretical computer science begins with the work of Turing. This early nontechnical article deals with the idea of artificial intelligence. Some of Turing's vocabulary is a little old-fashioned, but the ideas are important and relevant.

INTERNET RESOURCES

Mathematical ideas, especially ideas about numbers, are often subtle and expressed in an unfamiliar vocabulary. Without long periods of quiet reflection, mathematical concepts are often difficult to appreciate. This is exactly the type of environment one does not usually find on the Internet. To develop a real appreciation for mathematical thought, books are better. That said, the following Web sites are some good resources.

Boyne, Anne. Papers on History of Science. Available online. URL: http://nti.educa.rcanaria.es/penelope/uk_confboye.htm#particular. Accessed March 15, 2010. This is a very detailed and interesting paper devoted to the history of negative numbers. It is well worth reading.

Courtright Memorial Library, Otterbein College. History of Mathematics. Available online. URL: http://www.otterbein.edu/resources/library/libpages/subject/mathhis.htm. Accessed March 15, 2009. This is an excellent resource when planning additional reading.

Electronic Bookshelf. Available online. URL: http://www.dartmouth.edu/~mate/eBookshelf/trigonometry/index.html. Accessed April

23, 2010. Maintained by Dartmouth College. This site has links to many creative presentations on computer science, the history of mathematics, and mathematics. It also treats a number of other topics from a mathematical perspective.

Eric Weisstein's World of Mathematics. Available online. URL: http://mathworld.wolfram.com/. Accessed April 23, 2010. This site has brief overviews of a great many topics in mathematics. The level of presentation varies substantially from topic to topic.

Faber, Vance, Bonnie Yantis, Mike Hawrylycz, Nancy Casey, Mike Fellows, Mike Barnett, Gretchen Wissner. This is MEGA Mathematics! Available online. URL: http://www.c3.lanl.gov/mega-math. Accessed April 23, 2010. Maintained by the Los Alamos National Laboratories, one of the premier scientific establishments in the world, this site has a number of unusual offerings. See, especially, "Welcome to the Hotel Infinity!"

Gangolli, Ramesh. *Asian Contributions to Mathematics.* Available online. URL: http://www.pps.k12.or.us/depts-c/mc-me/be-as-ma.pdf. Accessed April 23, 2010. As its name implies, this well-written on-line book focuses on the history of mathematics in Asia and its effect on the world history of mathematics. It also includes information on the work of Asian Americans, a welcome contribution to the field.

Jabberwacky. Available online. URL: http://www.jabberwacky.com/. Accessed March 15, 2009. Alan Turing believed that a computer would be able to successfully imitate a human in conversation by the year 2000. This Web site has a computer that is programmed to chat with the user. See if you are fooled. (Be sure to read Turing's own thoughts on the subject in the article, "Can a Machine Think?" referenced in the "Original Sources" subsection of the "Further Resources" section of this volume.)

O'Connor, John L., and Edmund F. Robertson. The MacTutor History of Mathematics Archive. Available online. URL: http://www.gap.dcs.st-and.ac.uk/~history/index.html. Accessed April 23, 2010. This is a valuable resource for anyone interested in learning more about the history of mathematics. It contains an extraordinary collection of biographies of mathematicians in different cultures and times. In addition it provides information about the historical development of certain key mathematical ideas.

Podnieks, Karlis. Available online. URL: http://www.ltn. lv/~podnieks/. Accessed March 15, 2009. This is Professor Podnieks's home page. It contains links to articles by him and others on the foundations of mathematics. Especially recommended is the link to his article, "Are You a Platonist? Test Yourself."

PERIODICALS, THROUGH THE MAIL AND ONLINE

+Plus

URL: http://pass.maths.org.uk

A site with numerous interesting articles about all aspects of high school math. They send an email every few weeks to their subscribers to keep them informed about new articles at the site.

Pi in the Sky

http://www.pims.math.ca/pi/

Part of the Pacific Institute for the Mathematical Sciences, this high school mathematics magazine is available over the Internet.

Scientific American

415 Madison Avenue
New York, NY 10017

A serious and widely read monthly magazine, *Scientific American* regularly carries high-quality articles on mathematics and mathematically intensive branches of science. This is the one source of high-quality mathematical information that you will find at a newsstand.

INDEX

Italic page numbers indicate illustrations.
Page numbers followed by *c* indicate chronology entries.